Introduction to the Kinetics of Glow Discharges

Introduction to the Kinetics of Glow Discharges

Chengxun Yuan
Harbin Institute of Technology, China

Anatoly A Kudryavtsev
St. Petersburg State University, Russia and Harbin Institute of Technology, China

Vladimir I Demidov
West Virginia University, USA

Morgan & Claypool Publishers

ISBN   978-1-64327-060-9 (ebook)
ISBN   978-1-64327-057-9 (print)
ISBN   978-1-64327-058-6 (mobi)

DOI   10.1088/978-1-64327-060-9

Version: 20180801

IOP Concise Physics
ISSN 2053-2571 (online)
ISSN 2054-7307 (print)

A Morgan & Claypool publication as part of IOP Concise Physics
Published by Morgan & Claypool Publishers, 1210 Fifth Avenue, Suite 250, San Rafael, CA, 94901, USA

IOP Publishing, Temple Circus, Temple Way, Bristol BS1 6HG, UK

Dedicated to the memory of Professor Lev Tsendin (1937–2012), whose contributions to the study of gas discharge and plasma science were longstanding and far-reaching.

Contents

Preface

Electric glow discharges (glows) can be found almost everywhere, from atmospheric electricity to modern plasma technologies, and have long been the object of research. They are described in a number of monographs written by the leading scientists. Although those books have a well-deserved popularity, the fundamental foundations are mainly described on the basis of a fluid (hydrodynamic) approximation. In that approach the electron distribution function (EDF) depends on the local plasma parameters at a given point of space in a given time. However, under conditions of a gas-discharge plasma, the EDF is usually strongly non-equilibrium and the fluid approximation often cannot describe some important phenomena and processes even qualitatively.

The appropriate approach for describing such highly non-equilibrium systems and phenomena is physical kinetics. In recent years, work in this area has progressed rapidly and a self-consistent kinetic description of the simplest glow discharges has become possible. The information about that approach is scattered across many articles and reviews. The main purpose of this book is to provide simple illustrations of the basic physical mechanisms and principles that determine the properties of those categories of discharges and to enable readers to explore modern literature and successfully participate in scientific and technical progress.

Symbols and abbreviations

α	the first Townsend coefficient
γ	secondary electron yield
γ_a	apparent secondary electron yield
γ_A	secondary electron yield from atoms
γ_{eff}	the effective secondary electron yield
δ	double ratio of electron and atom masses
δ_1	small perturbation of a certain value
ε	total electron energy
ε_0	ionization degradation energy
ε_1	the first excitation potential of an atom
ε_c	ionization efficiency
ε_f	fast electron energy
ε_i	the first ionization potential of an atom
ε_m	energy of maximum F
ε_{st}	Stoletov constant
η	ionization energy parameter
$\ln \Lambda$	Coulomb logarithm
Λ	diffusion length
Λ_f	CF fast electrons range
λ	electron mean-free path
λ_{cs}	characteristic scale of decay of fast electrons
λ_i	ion mean-free path
λ_ε	elastic electron energy relaxation path
λ_ε^*	inelastic electron energy relaxation path
λ_E	electron energy relaxation length in electric field
$\lambda_{\widetilde{E}}$	microwave wave length
λ_{id}	ionization length
μ	electron multiplication coefficient
ν	electron collision frequency
ν^*	inelastic electron-atomic frequency
ν_i	direct ionization
	electron-atomic frequency
ν_{ia}	ion–atom collision frequency
σ_i	ionization cross-section
τ_ε	energy relaxation characteristic time
τ_{df}	electron diffusion time
τ_e	electron loss time
τ_i	ion drift time to anode
τ_{sn}	time for instability evolution in Townsend discharge
Φ	electron energy differential flux
Φ_{pl}	voltage drop across plasma
Φ_{sh}	voltage drop over sheath
Φ_w	voltage drop between plasma axis and wall
φ	plasma space potential
φ_0	anode potential
Ω	solid angle
Ω_1	frequency of perturbation

Ω_d	decrement of oscillations
Ω_i	increment of oscillations
ω	electromagnetic field frequency
$\mathscr{A}$	electron oscillation amplitude
A, B, C, D	constants from equations (2.4) and (2.5)
$\mathscr{C}$	capacitance of discharge-circuit system
b_e	electron mobility
b_i	ion mobility
D_e	electron free diffusion coefficient
D_a	electron ambipolar diffusion coefficient
D_ε	electron energy diffusion coefficient
d	cathode sheath thickness (also d_1, d_2)
E	electric field
E_x	electric field along discharge axis
E_r	radial electric field
$\mathscr{E}$	electromotive force, EMF
E_{dc}	direct current (dc) electric field
E_0	amplitude of ac electric field
E_{eff}	effective electric field
E_{cr}	critical electric field for runaway electrons
E_m	Stoletov constant corresponding electric field
e	electron charge
$F_0(w)$	local EDF
F	electron retarding force
f_0	isotropic part of EDF
f_1	directed part of EDF
f_{es}	$=j_{e0}/j_{emit}(0)$
$\mathscr{F}$	function connecting α/p and E/p (see equations (2.4) and (2.5))
I_r	radiation intensity
i	discharge current
$I(x)$	ionization rate
j_+, j_-	electron current to and from cathode
j_e	density of electron current
j_{ns}	current density on the cathode
j_n	normal current density
j_{emit}	current emitted from cathode
j_{eo}	primary electron current density in the near-cathode region
j_i	density of ion current
j_f	density flux of fast electrons
j	total current density, $j_e + j_i$
L	discharge gap length
L_{FDS}	length of FDS
L_{NG}	length of NG
L_{PC}	length of PC
$\mathscr{L}$	loss function
M	electron multiplication coefficient in discharge gap
m	electron mass
N	density of atoms
n	plasma density
n_e	density of electrons

n_i	density of ions
p	pressure
R	circuit resistance
R_d	discharge (tube) radius
S_c	cathode area
T^*	temperature of EDF fast part
T_e	electron temperature
T_0	gas temperature
t	time
V	electron speed
$\overline{V}$	average electron speed
u_i	ion drift speed
u_e	electron drift speed
U	voltage between cathode and anode
U_c	sheath voltage
U_{cn}	potential drop across cathode
U_b	breakdown voltage
U_{ns}	cathode potential drop
w	kinetic electron energy
x	axis along discharge
x_m	plasma density point of maximum
Z	ion charge number
Z_i	ionization frequency averaged over EDF
ac	alternative current
AF	anode fall
CF	cathode fall
dc	direct current
DS	dark space
EDF	electron distribution function
FDS	Faraday dark space
FR	field reversal
NG	negative glow
PC	positive column
PNG	plasma part of NG

Author biographies

Chengxun Yuan

Chengxun Yuan received his PhD degree in physics from Harbin Institute of Technology in 2011. He is now the Associate Professor at the Department of Physics, Harbin Institute of Technology, Harbin, China. His research interests include basic plasma physics, dusty plasma physics and the interaction of electromagnetic radiation with plasma. He has more than 80 publications in the plasma field.

Anatoly A Kudryavtsev

Anatoly A Kudryavtsev received the MS and PhD degrees in physics from the Leningrad State University, Leningrad, USSR (now St Petersburg, Russia), in 1976 and 1983, respectively. Since 1982, he has been with the St Petersburg State University, where now he is an Associate Professor in the Optics Department, and the Chair Professor at the Department of Physics, Harbin Institute of Technology. He is an expert in gas discharge physics and the kinetic theory of plasma physics, and the author of over 150 journal articles and numerous conference presentations.

Vladimir Demidov

Vladimir Demidov received his PhD degree from Leningrad State University (now St Petersburg State University), Leningrad, USSR (now St Petersburg, Russia), in 1981. He has been with West Virginia University (Morgantown, WV, USA) since 2003. He is now a Research Professor at the Department of Physics and Adjunct Professor at the Department of Mathematics. His primary interests are basic plasma physics, plasma diagnostics, and atomic and molecular physics. He has more than 150 publications and is the author of a monograph on probe methods of plasma diagnostics.

IOP Concise Physics

Introduction to the Kinetics of Glow Discharges

Chengxun Yuan, Anatoly A Kudryavtsev and Vladimir I Demidov

Chapter 1

Introduction

Electric discharges with a relatively low contribution of power and weak current are usually referred to as glow discharges or glows. The glows have different forms, are characterized by the frequency of the supply voltage, the size and the shape of the electrodes and discharge volumes, and the composition and pressure of gas fillings. The golden age of physics of gas discharges has been associated with the names of J J Thomson, Townsend, Langmuir, von Engel, Steinbeck, Druyvesteyn and others and has been marked by such fundamental achievements as discoveries of electrons and x-rays, development of spectral analysis, etc. Those achievements have established the foundation of modern physics. At present, in order to study the fundamental laws of nature, other methods are used. But the applied importance of processes in gas discharges has immeasurably increased. In this regard, there is an important need to be able to create gas-discharge devices with predefined properties and parameters. To do that, we need to have a deep understanding of the physical processes in the discharges.

Up to now considerable progress has been made in understanding the kinetics of discharges. It is now possible to obtain characteristics of the simplest types of discharges from 'first principles'. The most clearly kinetic phenomena have manifested in weak-current discharges of low and medium pressure. In such discharges, plasma is very far from thermodynamical equilibrium. The external energy in the discharge mainly goes primarily to electrons. Transferring it from electrons to heavy plasma particles is usually, for some reason, difficult and the removal of energy from heavy particles, especially at low pressures, occurs relatively quickly. As a result, in the discharges, the most non-equilibrium part is the electron component, so for its description the kinetic analysis is really necessary. Logical development of this area of physics inevitably leads to the fact that the description of the discharges becomes the kinetic one.

To understand physics and technology at the modern level, it is necessary to be well versed in this field. Recently, work in this direction has been developing

doi:10.1088/978-1-64327-060-9ch1

intensively, and although, in our opinion, only the first steps have been taken, a self-consistent kinetic description of the simplest glow discharges has became possible. However, this information has been scattered across numerous articles and reviews.

The purpose of this book is to illustrate the basic physical mechanisms and principles of the glows, enabling readers to study the modern literature and successfully participate in scientific and technical progress. There are a number of excellent textbooks and monographs on the physics and phenomenological description of gas discharges of various types. Among them are the book by Lieberman and Lichtenberg [1], a series of books by Raizer and co-workers [2–4], Biberman's book [5], the books of Chen [6], Smirnov [7] and others. We can also mention a recent review [8]. The nature of the present book (as well as books [9, 10], which are published in Russian only) and its main difference from the above publications are, first of all, in its deeper kinetic level of description of the glow discharges. That allows us to describe glows in a single and much more precise manner of the basic properties of the discharges. All of the the development of modern physics of discharges contributes to the deepening and development of a more accurate primarily kinetic description. This gives the possibility to adequately describe these objects quantitatively, which is extremely important for numerous technical applications. Reading this book assumes acquaintance with one of the above monographs, as well as some knowledge of charge particle kinetics in gas-discharge plasmas (for example, book [11] and reviews about non-local electron kinetics [12–14]) and mathematics at the level of a standard university course.

This book is structured as follows. The second chapter of the book discusses ionization in the electric fields. The electrons and ions may be generated in the gas volume by ionization, whose intensity strongly depends on the field. In the weak field there is no, or not enough, generation of charged particles to have self-sustained discharge. Townsend established experimentally the ionization relationships and introduced the key ionization characteristic (the first Townsend coefficient). The chapter concludes with a brief discussion of the differences of ionization in low and high electric fields, including the influence of runaway electrons. The ionization in the strong field may be essentially non-local and the concept of the first Townsend coefficient may not be applicable.

The third chapter is devoted to the description of the microwave breakdown, which is a more simple case with respect to the dc breakdown. In this case diffusion of electrons to the volume walls may be important. The breakdown behavior may depend on the wave frequency ω. Application to the discharge gap of even a small dc electric field may also change the breakdown properties.

In chapter 4 the description of the dc breakdown is provided, the second Townsend coefficient is introduced and the Paschen curve is discussed. It is demonstrated that the breakdown on the right-hand side of the Paschen curve is in good agreement with the local theory. At the same time, the left-hand side of the Paschen curve can occur due to essentially non-local processes, since in this case the breakdown occurs in strong electric fields. This chapter discusses the relationship and the difference between the breakdown of Townsend [15], formulated for the breakdown of a direct current, and the Brown's criterion [16] for the breakdown in microwave fields.

The general structure of a discharge between cold electrodes is described in chapter 5. Most discharge laboratory experiments have been conducted in long cylindrical tubes. In this section the IV-traces and axial structures are discussed. A typical discharge has a number of areas like the Faraday dark space (FDS), negative glow (NG), and positive column (PC). Some of them may be absent (like PC).

In chapter 6 we consider the Townsend discharge, in which the effect of the space charge is negligible, and conditions for the applicability of the traditional hydro-dynamic approximation are discussed. It turns out that for the discharge on the right branch of the Paschen curve we apply a local approach in which ionization is approximated using the first Townsend coefficient α, which depends on the local value of the electric field, while the effective coefficient for ion–electron emission γ_{eff} is substantially reduced due to kinetic effects. As the discharge current rises, the electric field distortion by the space charge begins to play a role and the Townsend discharge on the right branch of the Paschen curve becomes unstable. The develop-ment of this instability leads to the establishment of a mode of normal current density. In this mode, ionization becomes non-local. The situation on the left branch of the Paschen curve is different, The field is rather high and at mean free path an electron acquires considerable energy. As a result, the phenomenon of runaway electrons begins to play a substantial role. For its description a sequential kinetic analysis is required [17].

Chapter 7 is devoted to a description of a short (without positive column) dc discharge. Since the positive column is not an obligatory discharge structure, this allows us to consider only the most significant discharge phenomena (without which the discharge cannot exist), that is, those phenomena that occur at the electrodes. The cathode region of the discharge is essentially non-local. Attempts to describe it in the hydrodynamic approximation are not allowed to simulate all phenomena occurring in this region. Therefore, its description must be essentially kinetic. The anode region still does not have a comprehensive and detailed explanation.

The positive column of a glow dc discharge is the most studied plasma object. Its kinetic analysis is described in chapter 8. It turns out that the traditional hydro-dynamic approach cannot adequately describe many important characteristics of even this simplest object; only kinetic analysis allows us to explain these character-istics consistently. At low and medium pressure, the EDF in the PC is non-local, i.e. it does not factorize as a product of the electron density and a function of the velocity, the form of which is determined by the local value of the axial electric field. In the limiting case of low pressures there is a complete non-locality of the EDF and it depends only on the total energy ε (the sum of the kinetic energy w and the potential energy in the electric field $-e\varphi$).

Sections 4.3, 7.3, 7.4 and 7.5 have been written by V Demidov and A Kudryavtsev. The rest of the text has been written by A Kudryavtsev and C Yuan. Some important issues connected to glows have not been considered in this book. These are detailed electron kinetics, striations, discharges with hollow cathodes and some others which the authors intend to discuss in separate publications.

References

[1] Lieberman M and Lichtenberg A 2005 *Principles of Plasma Discharges and Materials Processing* (New York: Wiley)

[2] Raizer Y P 1991 *Gas Discharge Physics* (Springer: Berlin)

[3] Raizer Y P, Schneider M and Yatsenko N 1995 *Radio-frequency Capacitive Discharges* (Boca Raton, FL: CRC Press)

[4] Bazelyan E M and Raizer Y P 1998 *Spark Discharge* (Taylor Francis: London)

[5] Biberman L M, Vorobev V S and Yakubov I T 1987 *Kinetics of Non-equilibrium Low-Temperature Plasmas* (Springer: Berlin)

[6] Chen F 2016 *Introduction to Plasma Physics and Controlled Fusion* (Springer: New York)

[7] Smirnov B M 2015 *Theory of Gas Discharge Plasma* (Springer: Berlin)

[8] Gudmundsson J T and Hecimovic A 2017 Foundations of dc plasma sources *Plasma Sources Sci. Technol.* **26** 123001

[9] Golubovskii Y B, Kudryavtsev A A, Nekuchaev V O, Porohova I A and Tsendin L D 2004 *Electron Kinetics in Non-equilibrium Gas-discharge Plasma* (St Petersburg: SPbSU) [in Russian]

[10] Kudrayvtsev A A, Smirnov A S and Tsendin L D 2010 *Physics of Glow Discharge* (St Petersburg: Lan) [in Russian]

[11] Shkarofsky I P, Johnston T W and Bachynski M P 1966 *The Particle Kinetics of Plasmas* (North Miami Beach, FL: Addison-Wesley)

[12] Tsendin L D 1995 Electron kinetics in non-uniform glow discharge plasmas *Plasma Sources Sci. Technol.* **4** 200

[13] Tsendin L D 2010 Electron kinetics in non-uniform glow discharge plasmas *Phys.-Usp* **53** 133

[14] Kolobov V I and Godyak V A 1995 Nonlocal electron kinetics in collisional gas discharge plasmas *IEEE Trans. Plasma Sci.* **23**

[15] Townsend J S 1915 *Electricity in Gases* (Clarendon: Oxford)

[16] Brown S C 1956 *Handbuch der Physik* vol 22 (Heidelberg: Springer)
Brown S C 1951 *Proc. Inst. Radio Eng.* **39** 1493

[17] Babich L P 2003 *High-Energy Phenomena in Electric Discharges in Dense Gases: Theory, Experiment and Natural Phenomena* (Arlington, VA: Futurepast)

Chapter 2

Ionization in the electric field and the first Townsend coefficient

When a neutral gas is ionized by an electric field, it becomes conductive. The breakdown in the gas may ignite a stationary self-sustained discharge operating independently of the presence or strength of the external ionizer [1]. The electrons (and ions) are generated in the gas by ionization, whose intensity strongly depends on the field. Charged particles escape the discharge volume via diffusion and drift to the boundaries and electrodes, as well as via recombination and attachment in its bulk. These escape mechanisms, however, are not so much dependent on the field. The breakdown occurs only in the electric field exceeding a certain threshold value specific to the particular conditions. The criterion for a breakdown (see chapter 4) was formulated by Townsend as the rate equality of electron multiplication by impact ionization and their loss, which must hold true at any point of the discharge volume [2]. Since a gas breakdown may occur in an electromagnetic field of any frequency ω, the respective criterion is defined by the effective field $E_{\text{eff}} = (E_0 \nu)/(\nu^2 + \omega^2)^{1/2}$. At a given field E_{eff}, the breakdown is usually identified by comparing three characteristic lengths: the gap length L, the electron free path λ (or pressure p), and the electron oscillation amplitude $\mathcal{A} = (eE_0)/[m\omega(\nu^2 + \omega^2)^{1/2}]$. A change in the relationship among these parameters may modify the breakdown pattern radically [1, 3, 4]. Various scenarios of the breakdown behavior have been discussed in [1, 3, 5].

A breakdown is a very complex physical phenomenon. Its most important component is the initial electron avalanche developing in the gas due to the ionization by seed electrons under the action of an electric field. It is of little importance to the ionization whether electrons are generated during a regular avalanche drift in a dc field or they 'tramp about', oscillating at a high rate in an alternating field.

The key process in the production of charged particles is a direct ionization by electron impact, whose rate is characterized by frequency $\nu_i = N\langle V\sigma_i\rangle$ [6], where $\langle\rangle$ is the averaging over EDF, N is the density of atoms, V is the electron speed and σ_i is the electron impact ionization cross-section.

Townsend carried out experiments for determining the ionization characteristics and their interpretation [2]. In a sufficiently high field, every seed electron initiates an avalanche; this is illustrated schematically in figure 2.1 for a dc field. If the electrons drift at an average speed $u_e = b_e E$, their current density is $j_e = e n_e u_e$, and a convenient characteristic to describe the ionization evolution along the coordinate is the first Townsend coefficient, $\alpha = \nu_i / u_e = \nu_i / (b_e E)$, equal to the number of ionization events per electron per unit path length in the field E. The stationary electron balance equations can be written as

$$\partial j_e / \partial x = \alpha j_e, \quad \partial n_e / \partial x = \alpha n_e. \tag{2.1}$$

The ionization in a dc field grows exponentially (in an avalanche-like manner)

$$j_e(x)/j_e(x = 0) = n_e(x)/n_e(x = 0) = \exp(\alpha x) = M(x), \tag{2.2}$$

where $M(x)$ is the electron multiplication coefficient in the gap.

In Townsend's experiments, a chamber with plane-parallel electrodes with a variable gap length L could be filled with different gases under controllable pressure p. The current density j was uniform over the cross-section. A constant voltage U, which could also be varied, was applied between the cathode and the anode. The cathode was uniformly irradiated by ultraviolet light to knock out seed electrons from its surface, which induced the primary electron current j_{e0} in the near-cathode region. When the voltage was raised gradually from a very small value, the circuit current first rose and reached saturation when all emitted electrons hit the anode (figure 2.2).

The fast current rise was due to the onset of avalanche multiplication of electrons. A exponentially enhanced electron current was supplied to the anode, $j_e(L) = j_{e0} \exp(\alpha L) = j_{e0} M$ with $M = M(x = L)$. Since the ion current at the anode is zero, the same

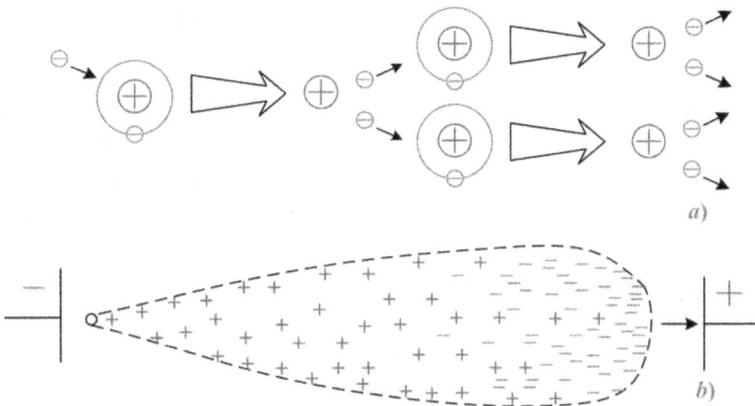

Figure 2.1. Schematic development of electron avalanche in the discharge gap between a negative cathode and positive anode: avalanche multiplication of electrons (a) and diffusive divergence of the avalanche from one electron (b). Electrons and ions are marked by pluses (+) and minuses (−), respectively.

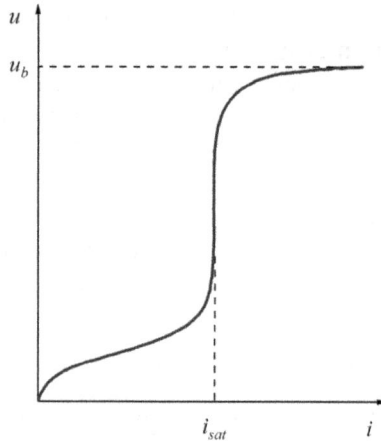

Figure 2.2. *IV*-trace $(i - u)$ of non-self-sustained discharge. u_b is the breakdown voltage and i_{sat} is the saturation current.

current goes through the entire circuit under stationary conditions and is the sum of the electron and ion currents. Every electron emitted by the cathode produces in the gap the $[\exp(\alpha L) - 1]$ number of positive ions which all reach the cathode. So the total cathode current is

$$j = j_e + j_i = j_{e0} + j_{e0}(\exp(\alpha L) - 1) = j_{e0}\exp(\alpha L) = j_{e0}M = j \qquad (2.3)$$

This current is not self-sustained: the discharge quenches when the external ionizer is turned off ($j_{e0} = 0$). By measuring the current at different gap lengths L but at constant U and p, one can find α as a function of $E/p = U/(pL)$ from the slope of the curve $\ln j/j_0 = \alpha L$. Up to the breakdown $(pL)_b$ value, the experimental relationships $\ln i(pL)$ at different E/p go along straight lines, as illustrated in figure 2.3. This fact was interpreted by Townsend as a check of the ionization relationship (2.2).

A similar approach was employed to find the ionization coefficients for many gases in various E/p ranges. The typical dependences are presented in figures 2.4 and 2.5 [5]. Reliable ways have also been suggested for finding the electron drift velocity [1, 7, 8]. Therefore, dc field experiments allow the measurement of α and u_e values, from which the ionization frequency $\nu_i = \alpha u_e$ can be found. These values can also be used when the ionization characteristics cannot be measured directly. For example, the $\alpha(E_{eff}/p)$ values for an effective field $E_{eff} = (E_0\nu)/(\nu^2 + \omega^2))^{1/2}$ can be used for an alternating field of frequency ω. Note that the EDF directly defines the ionization frequency ν_i, which is the primary characteristic of the ionization process.

Semi-empirical formulas are widely used for approximation of α/p, which is a function of E/p, $\alpha/p = \mathscr{F}(E/p)$

$$\alpha/p = A\exp(-Bp/E), \qquad (2.4)$$

$$\alpha/p = C\exp(-D(p/E)^{1/2}), \qquad (2.5)$$

The constants A, B, C, and D for various gases are given in table 2.1.

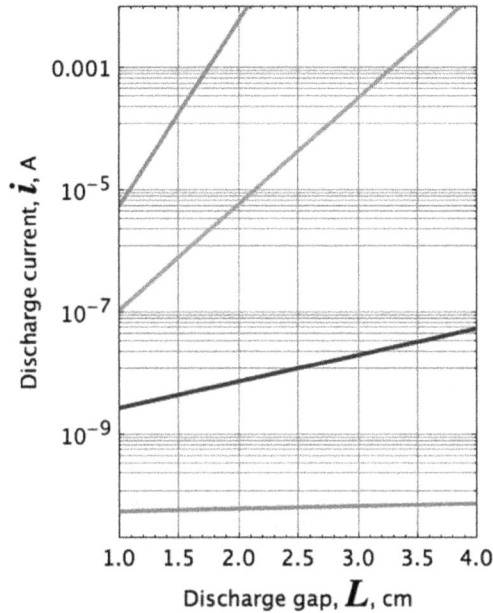

Figure 2.3. Exponential behavior of current rise through the discharge gap. E/p (in V/(cmTorr)) is equal to 36 (magenta), 33 (green), 30 (blue) and 26 (red).

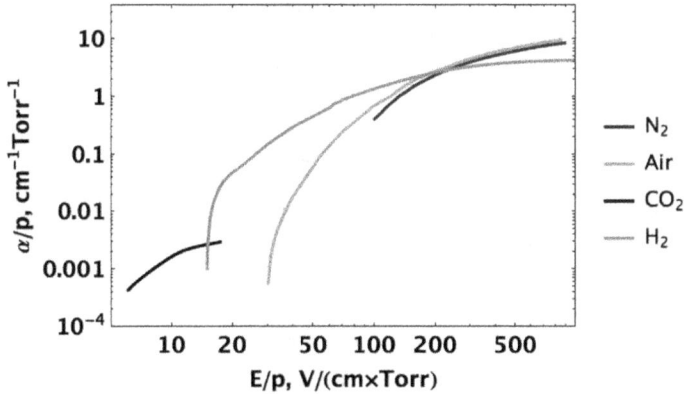

Figure 2.4. Ionization coefficient α/p in molecular gases.

By analogy with the spatial ionization parameter $\alpha = \nu_i/b_e E$, we can introduce the energy parameter $\eta(E/p) = \alpha/E = \nu_i/(b_e E^2)$ to describe the ionization efficiency in a uniform field. It is equal to the ratio of the ionization frequency to the Joule power per electron $(e b_e E^2)$ and gives the average number of ionization events performed by an electron which has passed the potential difference of one volt in its drift motion. From equation (2.2), we have $M = \exp(\eta U)$. It is sometimes more convenient to describe the ionization in a dc field by using η, not α. Traditionally, the coefficient α

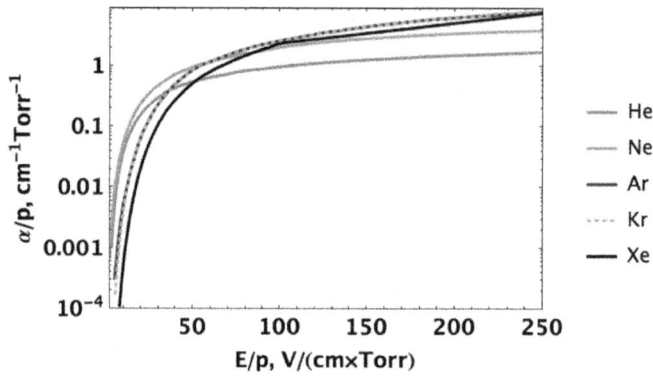

Figure 2.5. Ionization coefficient α/p in noble gases.

Table 2.1. The constants A, B, C, and D for various gases. From [1, 5].

Gas	A, (cm Torr)$^{-1}$	B, V (cm Torr)$^{-1}$	E/p, V (cm Torr)$^{-1}$
He	3	34	20–100
Ne	4	100	100–400
Ar	12	180	100–600
Kr	17	240	100–1000
Xe	26	350	200–800
Hg	20	370	200–600
H_2	5	130	150–600
N_2	8.8	375	27–200
Air	15	365	100–800
CO_2	20	466	500–1000
H_2O	13	290	150–1000

Gas	C, (cm Torr)$^{-1}$	D, V$^{0.5}$ (cm Torr)$^{-0.5}$	E/p, V (cm Torr)$^{-1}$
He	4.4	14	< 100
Ne	8.2	17	< 250
Ar	29.2	26.6	< 700
Kr	35.7	28.2	< 900
Xe	65.3	36.1	< 1200

is considered to be preferable because it can be easily measured in a dc field. The α value, however, is defined by the product of pressure and a function of E/p, so it varies with p at the same E/p values. In contrast, the coefficient η, or the ionization efficiency

$$\varepsilon_c(E/p) = e/\eta, \tag{2.6}$$

varies only with the E/p ratio and that may be more convenient.

It is clear from the above experimental data (figures 2.4 and 2.5) and from approximations (2.4) and (2.5) that the dependence of α/p (or η) on E/p consists of two very different segments. The dependence is very strong at small E/p values but weak at large E/p. The 'saturation' point of this relationship may be taken to be the maximum value $\eta = \eta_{max}$. Roughly, this point separates two limiting cases: the strong exponential dependence of α and η on the field and their approximately constant values.

The qualitative difference in the behavior of an electron gas in high and low fields can be explained as follows. If the energy of an electron injected into a low field is less than the ionization potential, then, before inducing ionization, the average electron must undergo many inelastic collisions and cover a distance much longer than $\lambda_E \sim \varepsilon_i/eE$, where it gains energy ε_i. Since the ionization probability is primarily determined by the exponentially small probability for the electron to overcome the energy range $\varepsilon_i - \varepsilon_1$, the ionization efficiency $\varepsilon_c = e/\eta$ is much greater than ε_i, and the ionization coefficients ν_i and α appear to be exponential functions of the local field E/p. The strong exponential dependence of α/p (see, equations (2.4) and (2.5)) on E/p means that the electron repeatedly gains energy from the field and loses it in elastic and inelastic collisions before performing the ionization, whereas the probability of gaining the ionization energy ε_i without losing it for excitation is exponentially small. The energies of the majority of electrons are either lower than the first ionization potential or only slightly higher. The elastic scattering at such energies dominates, and the EDF is close to isotropic.

In a smoothly non-uniform field (the typical characteristic scale of field non-uniformity in dc discharges is α^{-1}), the EDF is also local because the electron repeatedly gains and loses energy along the path $\lambda_E \sim \varepsilon_i/eE \gg \alpha^{-1}$ and eventually acquires equilibrium with the external field before it becomes available for multiplication, i.e. before gaining the energy ε_i along the path α^{-1}. For this reason, the ionization coefficients along the exponential section of the α/p on E/p curve are indeed the functions of the local field and can be calculated with a fairly good accuracy. A large number of such calculations have been made using the electron kinetic equation or direct Monte Carlo modeling. The ionization rate can be easily computed within a local approximation, using, for example, the well-known software packages Bolsig [9], Comsol Multiphysics [10] and others.

If the initial energy of an injected electron is much higher than ε_1 and ε_i, it is capable of starting the ionization immediately, irrespective of the local field strength. Before the EDF relaxes to its local field value, the electron may go a long distance and induce a noticeable *non-local ionization*, which will greatly exceed the local ionization at a particular local E/p ratio. For this reason in this case, one should not use the field-dependent parameters ν_i and α even for a low field. This happens, for example, when many electrons in a high field quickly acquire energies above ε_i. When going from a high field to a low field, they induce a considerable non-local ionization, and its description with the coefficients ν_i and α varying with the small local values of E/p would be incorrect. The EDF calculations for high fields are unreliable. The accuracy of direct numerical computations is often illusive because of the incomplete and often inaccurate cross-section data for elementary events, especially for their angular dependences; besides, such computations are time-consuming and do not provide a clear physical picture. In fact, this problem can be (although it has not been) solved for atomic hydrogen (two-body problem), whose cross-sections in elementary events are known fairly well.

A local description of the electron behavior in a high field is inapplicable because of the runaway. As an electron gains energy from a high field, it has a non-zero probability to go over to the runaway, when collisions have almost no effect on its motion, making it practically free. A medium of even an infinite thickness would then become transparent to such an electron. Note, that some electrons injected with a low energy in a strong uniform dc field move practically freely and become infinitely accelerated, experiencing no scattering or retardation. This reveals itself in every gas having a specific critical field value E/p of order of 100 V (cm Torr)$^{-1}$, above which an electron continuously gains energy in spite of its loss in inelastic collisions. This fact is known experimentally [13].

A very approximate description of the effect may be made as follows. This effect happens because the inelastic energetic electron energy loss per unit path or the loss function

$$\mathcal{L}(w) = F(w)/N \qquad (2.7)$$

has a maximum due to the fact that the respective cross-sections are limited and decrease with energy (see figure 2.6). This maximal value, F_m, normally lies in the vicinity of the energy $\varepsilon_m \sim 100$ eV (see table 2.2). It is assumed here that energetic electrons are retarded with thermal electrons in inter-electron collisions [11]. That may be described as an action of a retarding force F, which is

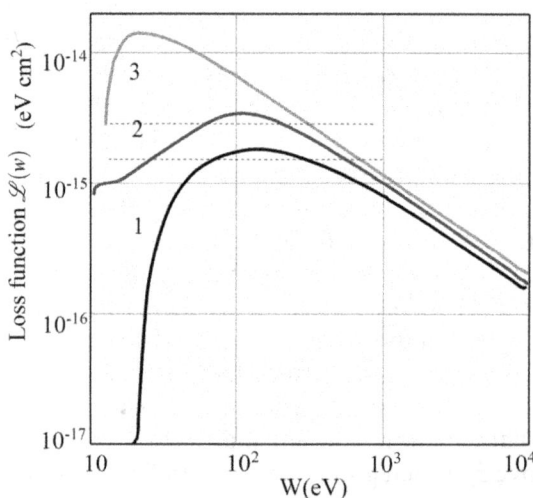

Figure 2.6. Energy-loss function $\mathcal{L}(w)$ for the fast electrons in molecular hydrogen (1) and helium (2). Calculation with Bethe–Bloch equation (2.9) for H_2 (3). Approximations 1.5×10^{-15} eV for He and 3×10^{-15} eV for H_2 (dashed lines). After [12].

Table 2.2. Critical electric field, E_{cr}/p (V (cm Torr)$^{-1}$), for runaway electrons at some gases. After [1, 5].

Cathode material	Gas									
	Air	Ar	He	H$_2$	Hg	Ne	N$_2$	O$_2$	CO	CO$_2$
Al	280	130	162	216	318	150	233	–	–	–
Au	285	130	165	247	–	158	233	–	–	–
Bi	272	136	137	240	–	–	210	–	–	–
C	–	–	–	240	475	–	–	–	526	–
Hg	–	–	142	–	340	–	226	–	–	—
K	180	64	59	94	–	68	170	–	484	460
Na	200	–	80	185	–	75	178	–	–	–
W	–	–	–	–	305	125	–	–	–	–
Zn	277	119	143	184	–	–	216	354	480	410

$$F(w) = \frac{2\pi n_e e^4}{w} \ln \Lambda \qquad (2.8)$$

for plasma [11] and similar

$$F(w) = \frac{\pi N Z e^4}{w} \ln \frac{4w}{I} \qquad (2.9)$$

for a neutral gas (Bethe–Bloch formula [13, 14]). For average ionization energy an expression $I = 10Z$ eV may be used [13, 14].

If eE is larger than F_m, from the equation for the energy variation along the electron path it follows that $E > E_{cr} = F_m/e$ and all electrons are runaways. The E_{cr}/p values for some gases are presented in table 2.2.

The uniform field strength for the transition to the runaway mode can, generally, be roughly estimated as the field, at which the dependence of α/p on E/p (see, equations (2.4) and (2.5)), reaches saturation (taking into account that such a description is less applicable with the field increasing). This saturation means that the electron is gaining the ionization energy practically without inelastic collisions. Since the ionization energy is the only characteristic atomic energy scale, at least in light gases, the electron will continue to gain energy with a large probability. The EDF and, hence, all local ionization parameters will be defined by the voltage the electron has passed, rather than by the local field. Therefore, it is more convenient here to use the energy parameter η, the more so as it tends to be constant at high energies. Physically, it is clear that the η value must be small (with large ε_c) for electrons with a low initial energy in a low field and become saturated at high fields. This obscure fact that the approximations (2.4) and (2.5) give a peak in the $\eta(E/p)$ curve is additional evidence that the ionization characteristics are independent of the local field in a high field, or in any field for high energy electrons. The approximation of the $\alpha(E/p)$ function (2.4) yields the η maximum

$$1/\eta_{max} = (2.72B)/A, \qquad (2.10)$$

Table 2.3. Ionization degradation energy and Stoletov constants for some gases. From [1, 5].

Gas	ε_0, eV	E/p, V $(\text{cm Torr})^{-1}$	$\varepsilon_{st} = (\alpha/E)_{min}^{-1}$, eV
He	46	50	83
Ne	37	100	66
Ar	26	200	45
Kr	24	200	42
Xe	22	300	38
Hg	–	200	80
H_2	36	140	70
N_2	37	350	75
O_2	33	–	–
CO_2	–	400	62
Air	36	365	66

which is attained in the field

$$(E/p)_m = B. \tag{2.11}$$

This point corresponds to the $(E/p)_m$ values, at which the tangential to the $\alpha/p(E/p)$ curve goes through the origin [1]. The energy $\varepsilon_{st} = e/\eta_{max}$, which has the meaning of the minimum potential difference to be passed by an electron in the field E_m prior to the ionization, was termed the *Stoletov constant* by Townsend. Since $\nu^* \gg \nu_i$, the Stoletov constant is as large as several ionization potentials: $\varepsilon_{st} = e/\eta_{max} = (3 - 4)\varepsilon_i$ (see table 2.3).

One can see that the value $2.72B/A$ (equation (2.10)) fits the experimental data well on ε_{st}. As the field rises further, there is only a slight ε_c increase (a slow fall of the ionizability η). Note one more time, that η has no real physical meaning for high fields.

Note that the ε_{st} value is close to, but not the same as, the energy of ionization degradation, ε_0, which a fast electron of energy ε_f spends for the creation of an electron–ion pair in full retardation. The number of ionization events in this case is $M = \varepsilon_f/\varepsilon_0$. For energies $\varepsilon_f > 1$ KeV, the ε_0 value practically does not change and is about two ionization potentials ($\varepsilon_0 \approx 2\varepsilon_i$). The values for the ionization degradation, ε_0, are also given in table 2.3.

Thus, to find the ionization characteristics in a field lower than $(E/p)_m$, one can use the calculations or experimental measurements of the Townsend coefficient and electron drift velocity in a dc field, which allow the ionization frequency to be found.

References

[1] Raizer Y P 1991 *Gas Discharge Physics* (Berlin: Springer)
[2] Townsend J S 1915 *Electricity in Gases* (Oxford: Clarendon)
[3] Francis G 1960 *Ionization Phenomena in Gases* (London: Butterworths Scientific)
[4] Brown S C 1959 *Elementary Processes in Gas Discharge Plasma* (Cambridge, MA: MIT Press)

[5] Golubovskii Y B, Kudryavtsev A A, Nekuchaev V O, Porohova I A and Tsendin L D 2004 *Electron Kinetics in Non-equilibrium Gas-discharge Plasma* (St Petersburg: SPbSU) [in Russian]

[6] McDaniel E W 1964 *Collision Phenomena in Ionized Gases* (New York: Wiley)

[7] Loeb L B 1939 *Fundamental Processes of Electrical Discharge in Gases* (New York: Wiley)

[8] Von Engel A 1955 *Ionized Gases* (Oxford: Clarendon)

[9] http://www.bolsig.laplace.univ-tlse.fr

[10] http://www.comsol.com

[11] Golant V E, Zhilinskii A P and Sakharov I E 1980 *Fundamentals of Plasma Physics* (New York: Wiley)

[12] Kolobov V I and Tsendin L D 1992 Analytic model of the cathode region of a short glow discharge in light gases *Phys. Rev.* A **46** 7837

[13] Pitaevskii L P and Lifshitz E M 1981 *Physical Kinetics* (Amsterdam: Elsevier)

[14] Landau L D and Lifshitz E M 2003 *Quantum Mechanics* (Amsterdam: Butterworths)

IOP Concise Physics

Introduction to the Kinetics of Glow Discharges

Chengxun Yuan, Anatoly A Kudryavtsev and Vladimir I Demidov

Chapter 3

Microwave breakdown

The spatial distribution of ionization sources and the dominant mode of electron removal (diffusion or directed drift in the field) vary with the field frequency range. The limiting cases are microwave and dc breakdowns. We will consider in this chapter the simpler case of microwave breakdown, when the field frequency is so high that the electric field is reversed before the electron displacement due to the drift becomes noticeable. Then the electron loss is due to diffusion and the emission by the walls and electrodes is unimportant.

In microwave breakdown, the electrons and ions oscillate fast in the field without touching the walls, so the interaction with a surface or emission by them are normally of no importance. The electron avalanches are localized and develop at each site independently. For example, at a typical microwave frequency of $f = 3$ GHz and field amplitude $E_0 = 500$ V cm^{-1}, the amplitude of free electron oscillations is $\mathscr{A} = eE_0/(m\omega^2) = 2.5 \times 10^{-3}$ cm, which is much smaller than the characteristic gap length L (from 1 to 10 cm). Since the electron oscillation amplitude is small, the electron loss in a collisional mode ($\lambda \ll L$) is defined by the diffusion to the walls. The electrons are deposited there and recombine with incoming ions [1].

3.1 Microwave breakdown due to diffusion

After averaging over the microwave field period, the condition for a breakdown has the form suggested in [2, 3]

$$D_e\Delta n_e + \nu_i n_e = 0 \tag{3.1}$$

We will further consider the plane-parallel geometry $(0, L)$. We can use zero boundary conditions in the first approximation in λ/L. From equation (3.1), we get a sinusoidal distribution of the electron density smoothly falling towards the walls

$$n_e(x) = n_{e0} \sin(x/\Lambda), \tag{3.2}$$

doi:10.1088/978-1-64327-060-9ch3

and the breakdown condition is found as the eigenvalue of the boundary problem (3.1) [2]

$$\lambda_{id} = \Lambda, \qquad \nu_i = 1/\tau_{df} \tag{3.3}$$

Here,

$$\lambda_{id} = \sqrt{D_e/\nu_i} \tag{3.4}$$

is the ionization length, or the electron diffusion path for the time between two ionizing collisions, and

$$\tau_{df} = \Lambda^2/D_e \tag{3.5}$$

is the characteristic time of free electron diffusion, where Λ is the characteristic diffusion length in the discharge gap equal to L/π in the plane-parallel geometry and to $R_d/2.405$ in the cylinder. This condition means that the ionization frequency ν_i must compensate for the diffusion loss, i.e. an electron must perform, on average, just one ionization event for its lifetime. In contrast to the condition for a stationary plasma (see reference [1]), the expression (3.3) uses free electron diffusion instead of ambipolar diffusion. For this reason, the breakdown fields are higher than the plasma maintenance fields, and the microwave discharge typically develops for a time of the order of the free electron diffusion time. For example, the coefficient D_e is $\approx 2 \times 10^6 \text{ cm}^2 \text{ s}^{-1}$, and the time of the electron diffusion to the wall is $\tau_{df} \approx 5 \times 10^{-7}$ s in helium at $p = 1$ Torr and the diffusion length $\Lambda = 1$ cm.

When making microwave breakdown experiments, researchers plot the mean square field $E = E_0/\sqrt{2}$ (E_0 is the oscillation amplitude) as a function of pressure. A typical dependence of thresholds of microwave breakdown on gas pressure is shown in figure 3.1. In order to find ν_i, it is necessary to know the EDF, which can be

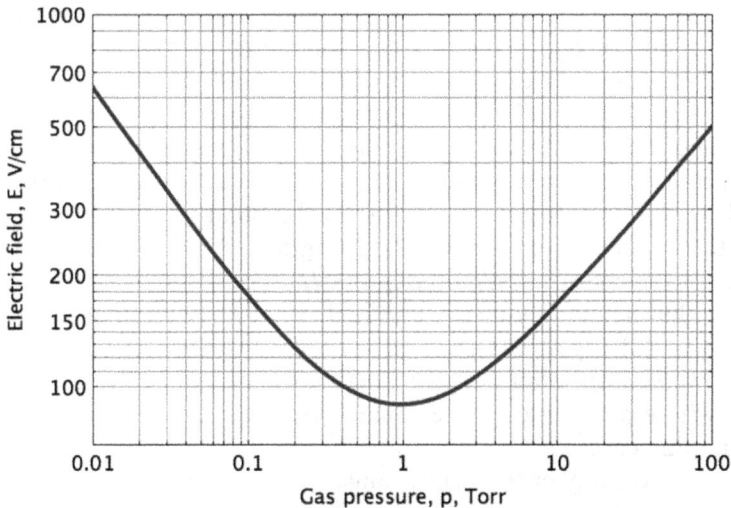

Figure 3.1. Typical dependence of measured thresholds of microwave breakdown (V cm^{-1}) on gas pressure (Torr).

calculated with reasonable accuracy [4, 5]. The substitution of the kinetic results on ionization frequency into the expression (3.3) gives an excellent agreement with experiments made in different gases under different conditions [6]. A reasonable fit is also achieved if one uses the results on ionization frequency obtained by the recalculation of the dc discharge data. An acceptable accuracy of inelastic collision and ionization frequencies in a microwave breakdown is possible within a 'black wall' approximation, which yields relatively simple analytical expressions for threshold fields [6].

Addition of readily ionizable seeds greatly decreases the breakdown field. This effect is especially appreciable in Penning gas mixtures, in which the seed is ionized both directly and by the excitation of long-living states of the buffer gas due to the Penning effect. This is especially evident for the so-called *Heg-gas,* a mixture of helium and a small amount of mercury, which served as a test object in microwave breakdown experiments [2]. Since elastic electron scattering is determined by buffer helium, the transport frequency ν can be taken to be energy independent. The Penning ionization of mercury leads, in turn, to the production of a new electron in each helium excitation event. There seem to be no inelastic energy losses for the excitation, and the ionization frequency coincides with the excitation frequency of the buffer gas (helium), which can be estimated in the black wall approximation (see [6]).

The major scalings can be derived from the analysis of the EDF formation [1, 6]. Since the diffusion coefficient along the energy axis, D_ε, in the kinetic equation for the isotropic EDF contains the effective field $E_{\text{eff}} = (E_0/\sqrt{2})[\nu^2/(\nu^2 + \omega^2)]^{1/2}$, the breakdown field varies only slightly with frequency at high pressures, when $\nu > \omega$ and $E_{\text{eff}} = E_0/\sqrt{2}$. The diffusion at high pressure is very slow, and a breakdown can be provided even by a low ionization rate. The electron energy balance is defined by elastic losses and electron temperature $T_e = eE_0\lambda_\varepsilon \simeq eE_0\lambda/\sqrt{\delta}$ [1]. Therefore, the breakdown field at low frequencies, $\omega \ll \nu$, must rise almost in proportion with pressure, as is observed experimentally. Assuming for the *Heg-gas* the average energy to be 1/3 of the *Hg* ionization potential, we estimate the breakdown field as $E = 2.4\,p$ V (cm Torr)$^{-1}$ [2]. At low pressures with $\nu \ll \omega$ and $E_{\text{eff}} = E\nu/\omega$, the breakdown field is proportional to the frequency at a given pressure. One has to raise the field strength in order to maintain the ionization rate and to compensate for the increasing loss due to diffusion, $D_e \sim 1/p$. So the breakdown field at low pressures varies with the frequency and the diffusion length Λ. After the necessary energy has been gained for the characteristic time $\tau_e \sim \varepsilon_1^2/D_e$, the *Heg-gas* becomes excited and ionized. For the elastic frequency $\nu = 2.4 \times 10^9 p$ (s^{-1}), the estimation yields $E = 1300/(p\Lambda\lambda_{\tilde{E}})$ [2] (figure 3.1), where $\lambda_{\tilde{E}}$ is the wavelength of the applied field. Hence, we observe a minimum in the breakdown curves in the (E, p)-coordinates (figure 3.1), whose position can be estimated from the expression $\nu \approx \omega$ separating the above limiting cases [2, 6].

This is true for the higher frequency ranges. For example, when a gas breakdown is produced by a high power laser [1], the relationship between the breakdown field and the gas pressure is the same as in a microwave breakdown.

In electronegative gases, the breakdown field rises considerably [2, 6] because of the additional electron loss due to attachment. The details of these phenomena have been described in [2, 6].

Thus, a microwave breakdown, for which the EDF calculations and the breakdown field measurements can be made with high accuracy, may serve as a convenient object for testing theories and experimental data. As the frequency of the electromagnetic field decreases, the electron oscillation amplitude becomes larger, reaching the value $\mathscr{A} = eE_0/(m\nu\omega)$ at $\omega < \nu$. Since the drift velocity is higher than the diffusion rate at $\lambda_e < \mathscr{A}$, the electrons which are at a distance $|x| < \mathscr{A}$ from the boundaries drift almost immediately to the walls and escape. As long as the condition $\mathscr{A} < L$ holds, the electron lifetime is defined by their diffusion through the discharge $(L - \mathscr{A})$, and the breakdown criterion coincides with equations (3.3) with L substituted by $\Lambda_{rf} = (L - \mathscr{A})/\pi$. As ω decreases and $\mathscr{A}$ approaches L, a shorter escape time will result in a higher breakdown field with a subsequent transition to a dc breakdown at $\mathscr{A} > L$.

3.2 Microwave breakdown in the presence of a low dc field

The next step towards a better understanding of the microwave breakdown physics was made owing to the experiments by Brown [2, 3] with an additional low dc field E_{dc}. In this case, electrons acquire definite directions of motion: the drift and diffusion fluxes travel in the same direction at the anode and in the opposite directions at the cathode. The stationarity condition requires that the electron escape should be compensated by ionization at each point in the discharge gap, so the cathode region becomes a 'weak point': there is the escape but the compensation is very small because the ionization (proportional to n_e) is actually zero. With the drift component of the velocity $b_e E_{dc} n_e$ in the dc field, the breakdown criterion (3.1) takes the form

$$D_e \nabla^2 n_e + b_e E_{dc} \nabla n_e + \nu_i n_e = 0. \tag{3.6}$$

By superimposing the zero boundary conditions on (3.6), we obtain the electron density profile [2]

$$n_e(x) = C_1 \exp(x/2\lambda_e)\sin(x/\Lambda), \tag{3.7}$$

where $\lambda_e = T_e/eE_{dc}$ is the energy relaxation length of an electron in the dc field. One can see in equation (3.7) a shift of the electron density profile towards the anode with decreasing λ_e due to the additional drift escape of electrons. This leads to a faster diffusion escape and, hence, to a higher ionization frequency, which is the eigenvalue of the problem for a fixed discharge gap. Mathematically, this means a shorter effective diffusion length λ_{id} under breakdown conditions (3.3)

$$1/\lambda_{id}^2 = D/\nu_i = 1/\Lambda_{dc}^2 = 1/\Lambda^2 + 1/4\lambda_e^2. \tag{3.8}$$

In this equation, the dc field strength enters into both the escape (the term with λ_e describing the enhanced escape) and the production (the expression for ν_i defined by the squared total field in equation (3.4). If the dc field is low, $\Lambda \ll \lambda_e$, its effect on the electron multiplication can be neglected and it is defined only by the microwave field

E_{ac}. The linearization of equation (3.8) and the substitution of $\nu_i(E_{ac})$ from equation (2.4) yield a quadratic dependence of the relative breakdown field E_{ac} rise on the small additional field E_{dc}. This dependence was experimentally observed in [2, 7]. Thus, in a low dc field, the microwave field amplitude must be increased to compensate for the electron drift escape in addition to the diffusion escape. The dc and microwave fields perform different functions here: the dc field stimulates an additional electron escape due to their drift to the cathode, whereas the microwave field provides the necessary ionization rate at any point in the discharge gap.

This breakdown mechanism does not involve ions and is entirely independent of the surface processes. In principle, a limiting case of this breakdown is possible in a high dc field, even with the microwave field turned off. When there is no electron emission from the cathode, the bottleneck region at the cathode produced by the drift escape to the anode can be compensated only by the diffusion back to the cathode. The limiting transition to the dc breakdown in equation (3.8) yields the relation $\lambda_{id} = 2\lambda_\varepsilon$ corresponding to

$$4\alpha\lambda_\varepsilon = 1 \quad \text{or} \quad \eta = e/(4T_e). \tag{3.9}$$

Even the minimum value of $1/\eta$ (Stoletov constant) is as large as $\varepsilon_{st} = e/\eta_{max} = (3 - 4)\varepsilon_i$ (see table 2.3), the relation (3.9) requires a high electron temperature, above the gas ionization potential $T_e > \varepsilon_i$. In other words, in order to fulfill the condition (3.9), the electron must perform an ionizing collision for a very short time of the order of $T_e/(b_eE^2)$, along a short path of about λ_ε, necessary for the drift velocity to be established. This requires unrealistically high fields. Equation (2.4) yields $E/p = B/\ln(4T_eAp/eE)$. This high field lies far above the Stoletov point, where this analysis has no sense at all.

In reality, however, other mechanisms may come into play even at a lower dc field. It is clear from the relation (3.7) and figure 3.1 that the electron density at the cathode drops sharply when the dc field is increased. A retarding field cools electrons, so the electrons moving against the field become cooler and there is no ionization in this field. Then even a slight electron emission by the cathode may appreciably change the value and direction of the electron flux, and the bottleneck at the cathode will disappear due to the cathode emission. The electron flux will reverse its sign and move away from the cathode such that the electrons will be heated by the accelerating field and multiply effectively.

References

[1] Raizer Y P 1991 *Gas Discharge Physics* (Berlin: Springer)
[2] Brown S C 1956 *Handbuch der Physik* vol 22 (Heidelberg: Springer)
      Brown S C 1951 High-frequency gas-discharge breakdown *Proc. Inst. Radio Eng.* **39** 1493–501
[3] Brown S C 1959 *Elementary Processes in Gas Discharge Plasma* (Cambridge, MA: MIT Press)
[4] http://www.bolsig.laplace.univ-tlse.fr
[5] http://www.comsol.com
[6] MacDonald A D 1966 *Microwave Breakdown in Gases* (New York: Wiley)
[7] Francis G 1960 *Ionization Phenomena in Gases* (London: Butterworths Scientific)

Chapter 4

Breakdown and self-sustained discharge ignition in a uniform dc field: the Paschen curves

As we discussed in chapter 3, the breakdown in microwave electric fields generally does not depend on the processes at discharge volume boundaries. The situation is different in the dc breakdown. In this case, the processes at the cathode are extremely important. The dc breakdown has been investigated by Townsend and is discussed below.

4.1 The secondary cathode emission effect on the current in a non-self-sustained discharge: the second Townsend coefficient

In Townsend's experiments discussed in chapter 2, the experimental points in the curve for $\ln j/j_0 = \alpha L$ against distance L at $E = U/L = $ const for a low field follow a straight line, whose slope can give the $\alpha(E/p)$ values. When the field and/or the gap length are increased, the curve deviates from a linear function (see figure 4.1), indicating that the current rises faster than in equation (2.1). In an attempt to understand this discrepancy, Townsend introduced the second ionization coefficient γ to account for the secondary effects on the cathode. For instance, when ions hit the cathode, they can knock out electrons with the efficiency

$$\gamma = j_{\text{emit}}(0)/j_i(0). \tag{4.1}$$

When interpreting the experimental data, one should bear in mind the important fact emphasized in [1, 2]. Equation (2.1), which only considers the drift of electrons, becomes invalid for the cathode region even in a uniform dc field because some of the electrons emitted by the cathode are backscattered without producing an ionizing collision. If equation (2.1) is used to describe the spatial electron current rise, the drift current entering the gap, $j_{e0} = n_{e0}u_e$ in equation (2.3), makes up only the fraction

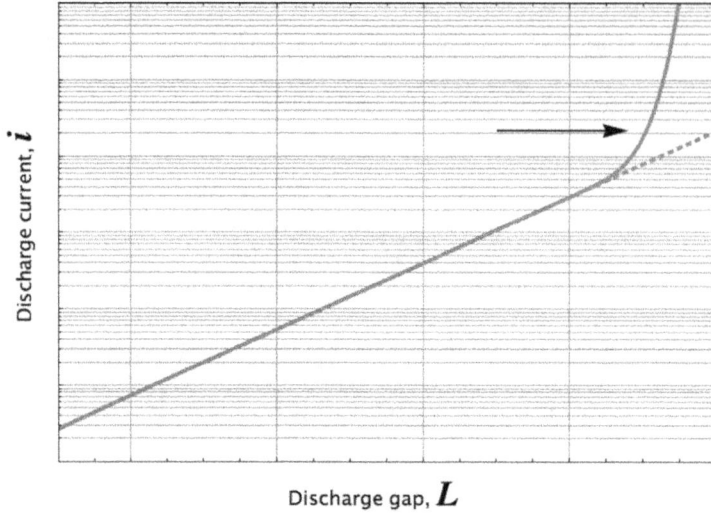

Figure 4.1. Typical influence of the secondary electron emission on increasing current in the discharge gap L.

$$f_{es} = j_{e0}/j_{emit}(0) \tag{4.2}$$

of the emission current $j_e(0)$ from the cathode surface. Factor f_{es}, which describes the fractional escape of electrons emitted by the cathode into the gas in [1, 2], was estimated on the assumption that the drift current j_{e0} to the anode was the difference between the emission current and the current back to the cathode. By equating the latter to the random electron current along the last free path length, we have

$$j_{e0} = n_{e0}u_e = j_{emit}(0) - n_e(\lambda)\bar{V}/4. \tag{4.3}$$

Assuming the electron density variation along the path to be small, $n_{e0} \approx n_e(\lambda)$, the factor f_{es} was estimated from equation (4.1) in [1, 2] to be

$$f_{es} = j_{e0}/j_{emit}(0) = 1/(1 + \bar{V}/4u_e). \tag{4.4}$$

It is clear from equation (4.4) that $f_{es} \ll 1$, because we have $u_e \ll \bar{V}$ [4]. This means that a large fraction of electrons must return to the cathode when the EDF is close to isotropic. The expression (4.3) has a simple physical meaning. The drift flux $n_e u_e$ is produced at the cathode but what goes back to the cathode is the random flux $n_e \bar{V}/4$; as long as $u_e \ll \bar{V}$, practically all the electrons go back to the cathode before they are 'picked up' and carried away by the field to the anode. The estimate from equation (4.3) shows that the emission current $j_e(0)$ must slightly exceed the random current to the cathode for the resulting electron current to be directed away from the cathode.

Because the $\bar{V}/u_e$ ratio varies with the field, the effective coefficient γ_{eff} in the boundary condition for the cathode in Townsend's model also varies with E/p. It is no longer a surface characteristic and it differs from the true γ by a factor of f_{es}

$$\gamma_{eff} = f_{es}\gamma. \tag{4.5}$$

Therefore, the full electron flux picked up by the field and carried away from the cathode is $j_{e0} = f_{es}j_0 + \gamma_{eff}j_i(0)$, where j_0 is the seed photocurrent from an external source. Using $j_i(0) = j_{e0}(e^{\alpha L} - 1)$, we find that the electron current across the anode, equal to the total current, is

$$j = j_{e0}e^{\alpha L} = j_0 f_{es}e^{\alpha L}[1 - \gamma_{eff}(e^{\alpha L} - 1)]^{-1}. \tag{4.6}$$

Its value is larger than that given by equation (2.3) due to the term between the square brackets, which becomes appreciable at large αL values. This expression agrees well with the experimental data at small γ. The respective dependences of the stationary circuit current on L in a dc field are presented in figure 4.1 and can be used to calculate both coefficients, α and γ_{eff}. Vast experimental information on the dependence of $\ln j/j_0$ on L can be found in [1, 4–8]. Figure 4.2 shows the γ_{eff} values for argon, obtained in this way by different researchers for cathodes made from different materials [9].

In was later shown in numerous experiments that the coefficient γ_{eff} is defined not only by ion–electron emission but also by other secondary processes. In addition to the ion collisions with the surface, an essential contribution is made by metastable and fast atoms in the ground state, photons, and other species. The observable $j(L)$ dependences do not allow the separation of this or that contribution, because it is not always possible to identify which of the mechanisms is dominant in an actual discharge. So one commonly uses the total emission coefficient γ_a, or the so called 'apparent' secondary electron emission yield, i.e. the yield per bombarding ion, including all secondary effects. The apparent yield varies in a wide range with the

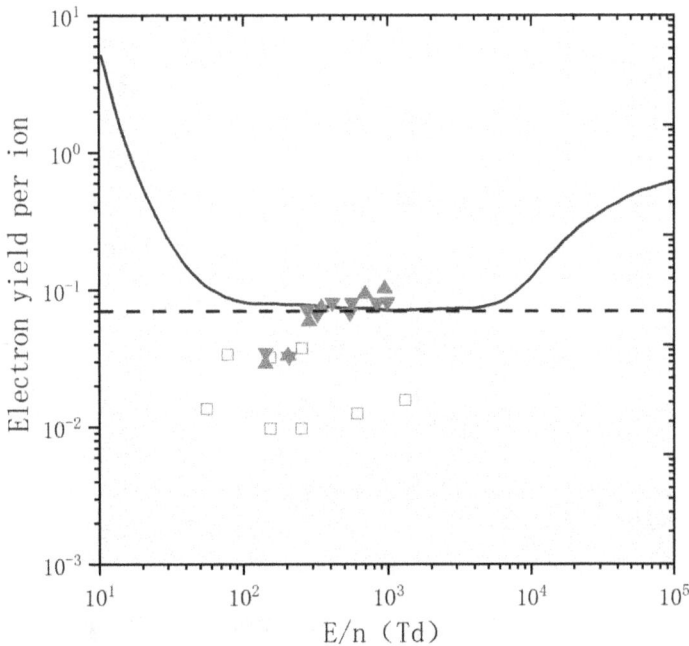

Figure 4.2. Values of γ_{eff} obtained for different pure surfaces. After [9].

gas, electrode material, applied field, current, etc; usually $\gamma \sim 10^{-2}$. The available data on the yield and the contributions of the various partial constituents to it are incomplete and often controversial. The yield values used by the authors in their calculations often differ considerably from the γ measured on pure surfaces in high vacuum [9].

4.2 The Townsend condition for self-sustained dc discharge ignition

The expression (4.6) was derived on the assumptions that the yield γ was small and that the electron multiplication coefficient $\mu = \gamma_{\text{eff}}(\exp(\alpha L) - 1)$ was smaller than unity. The denominator in equation (4.6) is positive and less than unity; therefore, the circuit current flows only if there is emission from the cathode. The discharge will remain non-self-sustained (the circuit current will also be zero at $j_{e0} = 0$). However, if the denominator in equation (4.6) becomes zero, this formally means the uncertainty $0/0$ at $j_{e0} = 0$. Physically, the circuit current will be observed even in the absence of an external electron source. The expression (4.6) was derived by Townsend in 1902 to explain the condition for the ignition of a self-sustained discharge. According to Townsend, it is an infinite current rise at $\mu \to 1$ that is the necessary condition for a breakdown and ignition of such a discharge

$$\mu = \gamma_{\text{eff}}(\exp(\alpha L) - 1) = 1, \quad \alpha L = \ln(1/\gamma_{\text{eff}} + 1). \tag{4.7}$$

This formula means that the discharge current may have a large value even without an external ionizer initiating the cathode emission. One primary electron from the cathode produces as many as $[\exp(aL) - 1]$ ions which, in turn, generate $\gamma_{\text{eff}}(\exp(\alpha L) - 1) = 1$ secondary electron on the cathode. Therefore, the Townsend criterion (4.7) defines the origin of a positive feedback when a primary electron generates a secondary electron capable of continuing this process. The equality of the cathode and anode currents

$$j = j_e(0) + j_e(0)/\gamma = j(L) = j_e(L) = j_e(0)\exp(\alpha L) \tag{4.8}$$

immediately yields the breakdown condition (4.7).

The field in a non-plane-parallel geometry is non-uniform, so equation (4.7) should be replaced by

$$M = \exp \int_0^L \alpha(E(x))dx. \tag{4.9}$$

Equation (4.7) suggests that the ionization is described by the local coefficient $\alpha(E/p)$, which is invalid at large E/p ratios, as was mentioned in chapter 2. The breakdown condition (4.7) should generally be written with the electron multiplication coefficient in the discharge gap M (equation (2.3))

$$\mu = \gamma_{\text{eff}}(M - 1) = 1. \tag{4.10}$$

The criterion (4.10) does not employ the assumption of the local dependence of $\alpha(E/p)$.

The effective coefficients $\gamma_{eff}(E/p)$ for various gases and cathode materials were tabulated in [1, 4, 8], using the Townsend formula (4.7) and experimental data on $\alpha(E/p)$ and the breakdown field. The representative values in figure 4.2 show that γ_{eff} may strongly depend on E/p. A more detailed study of the apparent secondary electron emission yield was made in [9] for argon, together with the data analysis of pure surfaces and those subjected to oxidation and other contaminating procedures (further, 'dirty' surfaces).

An analysis was made of the measurements of electron emission by ion and atom beam bombardment of pure and dirty metallic surfaces. It was found that the electron yield for a clean metallic surface in pure Ar was close to that for a clean metal in high vacuum at Ar^+ ion energy above 0.5 eV, which corresponds to $E/N > 250$ Td. However, analysis of the available data has shown that the electron yields for dirty metals differ greatly from those for clean metals and pure argon. Figure 4.3 gives the experimental electron yields for Ar^+ ions and Ar atoms incident on various clean metals as a function of the particle energy. One can see that the electron yield per ion is nearly independent of the ion energy below ~500 eV for most metals. At the ion energy of ~1 keV, the electron yield per ion is ~0.1 for most metals. In contrast, the yield per fast atom varies with energy, with an effective threshold at ~500 eV.

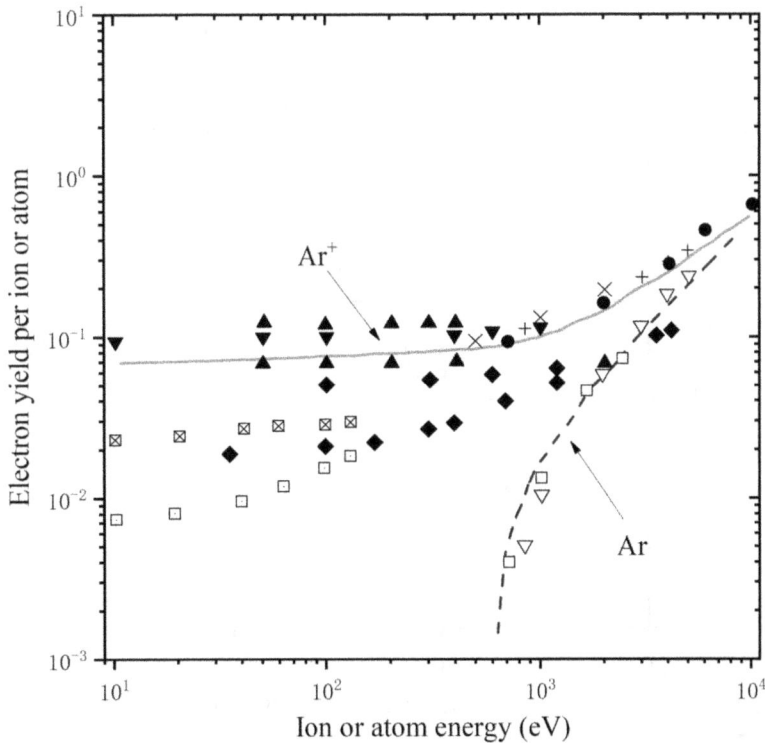

Figure 4.3. Electron yields for Ar^+ and Ar beams incident on various (W, Mo, Au, Cu, Pt and Ta) clean metal surfaces versus particle energy. The solid symbols are for Ar^+ and the open symbols are for Ar. After [9].

At high energies, the yield per fast atom approaches that for Ar^+ ions. Figure 4.4 presents the experimental electron yields γ for Ar^+ ions and γ_A for Ar atoms incident on metal surfaces with varying degrees of exposure to oxygen, water, ambient gas, or to unspecified contamination. The authors of [9] refer to these surfaces as 'dirty' (the terms 'practical' or 'laboratory' surfaces are also used in the literature). Above ~500 eV, the differences in the yields among metals are small compared to those for clean surfaces. The solid curves in figure 4.4 show the fits to the experimental beam data used in [9] compared with the swarm data. The dashed curves are averages over the data of figure 4.4 and show large changes in the yield that typically occur when a clean surface becomes oxidized or otherwise contaminated. The measurements for Ar^+ (open circles) in figure 4.4 indicate that there is more than a two-order spread in the yields at low energies (< 150 eV). Some of the low energy data show relatively large yields, while other data show very small yields.

Many of the Ar^+ data for energies below 100 eV suggest a relatively weak dependence on the ion energy characteristic of the potential ion–electron ejection. The measured electron yields per fast Ar atom, γ_A, indicated by solid circles in figure 4.4 have much the same energy dependence as those for fast Ar atoms incident on a clean metal but are shifted downward by about a factor of 10, such that the yields at a given energy are much larger. The authors of [9] made a thorough analysis

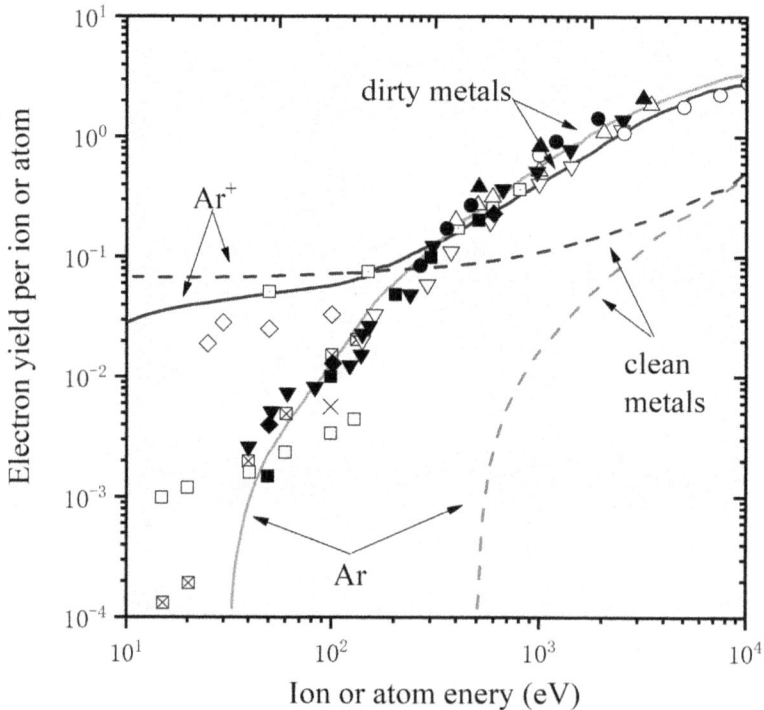

Figure 4.4. Electron yields for Ar^+ and Ar beams incident on various dirty metal surfaces versus particle energy. The open symbols are for Ar^+ and the solid symbols are for Ar. After [9].

of the available measurements of effective yields per ion, γ_{eff}, in pre-breakdown, breakdown and low current discharges. They made an attempt to separate the contributions of the various electron yields: those due to ions, metastable and fast atoms, and photoionization. An attempt was also made to identify the contribution of the electron backscattering to the cathode. Unfortunately, this kind of procedure for obtaining the electron yield from experimental data on each gas–metal pair is very time-consuming.

4.3 The ignition potential and the Paschen curves

If the coefficients α/p and γ_{eff} are the functions of E/p, then the breakdown condition equation (4.7) in a uniform field gives the functional dependence of the breakdown voltage U_b on (pL), known as the Paschen law [4]. This law was established experimentally by Paschen in 1889. The $U_b(pL)$ curves still bear his name. They are illustrated in figure 4.5 for various gases. When the state of the cathode surface changes, say, because of contamination during the cathode cleaning or formation of films, the ignition voltage may even become non-stationary. For this reason, the data of earlier experiments made with poorly cleaned cathodes differ from those of more recent studies using vacuum technologies. The condition (4.7) with known α and γ_{eff} defines the breakdown voltage U_b as a function of the gap length L. It follows from equation (4.7) for typical γ_{eff} values of $\sim 10^{-3}$–10^{-1} that an electron must perform from 3 to 10 multiplications along the gap length. In order to find explicit analytical dependences of the breakdown field on the gas, cathode material, pressure, and gap length, researchers often use equation (2.4) for $\alpha(E/p)$. By substituting it into equation (4.7), we get [4]

$$U_b = \frac{B(pL)}{\ln(pL) + C}, \qquad \frac{E_b}{p} = \frac{B}{\ln(pL) + C}, \qquad C = \ln \frac{A}{\ln(1/\gamma_{\text{eff}} + 1)}. \qquad (4.11)$$

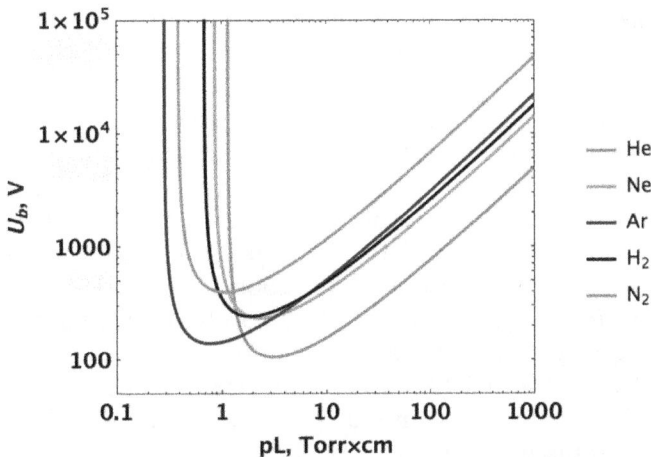

Figure 4.5. Paschen curves for various gases. After [10].

The calculation of U_b from equations (4.11) with the experimental constants A and B from table 2.1 gives a reasonable agreement with experiment. In practice, there are minimum voltages, at which the gap breakdown occurs. According to equation (4.11), we have at the minimum [4]

$$(pL)_m = \frac{2.72}{A} \ln\left(\frac{1}{\gamma_{\text{eff}}} + 1\right),$$

$$(E_b/p)_m = B, \qquad\qquad\qquad (4.12)$$

$$(U_b)_m = \frac{2.72B}{A} \ln\left(\frac{1}{\gamma_{\text{eff}}} + 1\right).$$

Clearly, the lowest ignition potentials should be expected for gases and cathodes with small B/A ratios and large γ values.

The general pattern of $(U_b)_m$ and $(pL)_m$ variations in different gases agrees with equations (4.12). The physical processes dominant in the right and left branches (relative to $(pL)_m$) of the Paschen curves differ considerably. At large pL values (the right branch), the breakdown voltage rises almost linearly with pL. This follows from the fact that the right branch corresponds to the exponential dependence of the ionization coefficient α/p on E/p, such that E/p remains nearly constant. Even a small variation in E/p changes the electron multiplication in the gap considerably. At small pL values (the left branch), however, the voltage U_b rises with a decrease in the number of collisions. Therefore, there is a minimum ignition voltage and the respective critical gap length $(pL)_m$. This value lies at the Stoletov point equation (2.10), at which the electron ionizability is largest, so the breakdown occurs most readily.

In contrast to $(U_b)_m$ and $(pL)_m$, the $(E/p)_m$ value given by equation (2.11) is independent of the cathode material (or γ_{eff}), as is observed experimentally. The saturation of the $\alpha/p(E/p)$ curve at large E/p ratios manifests formally as an infinite breakdown field rise with decreasing pL. The applicability of the concept of $\alpha(E/p)$ is limited by the runaway at large E/p. The α value is proportional to the gas pressure p, so the lower the pressure, the smaller the α value necessary for the transition to a continuous electron acceleration. The field at the start of the left branch is close to the runaway criterion $eE_m \sim F_m$ (see chapter 2) [11], so formulas of the type equations (4.11) and (4.12) do not have much physical meaning on the left from the minimum.

Therefore, it would be more reasonable to describe the electron multiplication using the concept of the energy for the electron–ion pair production, ε_c (equation (2.6)). In this approximation, the multiplication coefficient can be estimated as $M = eU_b/\varepsilon_{\text{st}}$. At smaller pL, the electron 'has no room' for multiplication since its ionization path is longer than the gap length, $1/\alpha > L$. In other words, the time of flight of a fast electron through the gap is insufficient for 'spending' all of its energy. Besides, some of the electrons may not only escape but also be reflected by the walls to be re-involved in the multiplication process.

For this reason, the calculation of the left branches of the Paschen curves require a detailed analysis of the electron interaction with boundaries and depends on the anode material. This may be associated with the fact that the gamma-electrons reflected by the anode have a sufficiently high energy to cover a longer distance in a retarding field and to ionize the gas additionally. Figure (4.6) shows the experimental data for the left branch in helium for various anode materials [12]. One can see that the curves are anomalous (non-monotonic). Similar behavior exhibits Paschen curves for mercury [13]. This is usually attributed to the peak in the energy dependence of the ionization cross sections or to the competition among various factors, in particular, to the contribution of secondary electron emission by neutral atoms. In the experiments [12], an automated system for measurement of Paschen curves, shown in figure 4.7 (left), was used. A wineglass discharge tube is shown in figure 4.7 (right). The discharge tube had two flat, disk-shaped electrodes facing each other at a variable distance from 2 to 20 mm. Most experiments were conducted at distance of 5 mm between the electrodes. The cathode was made from copper and was 50 mm in diameter. The anode has the same diameter, but was made from different materials. In this work, experiments were conducted with copper, aluminum, stainless steel, graphite, platinum-plated aluminum, and gold-plated aluminum anodes. The experiments were conducted in two regimes. In the first regime, for a certain pL, the voltage between the cathode and anode was slowly increased from zero to breakdown voltage. In the second regime, for a certain voltage between the cathode and anode, gas pressure was slowly increased from a value without

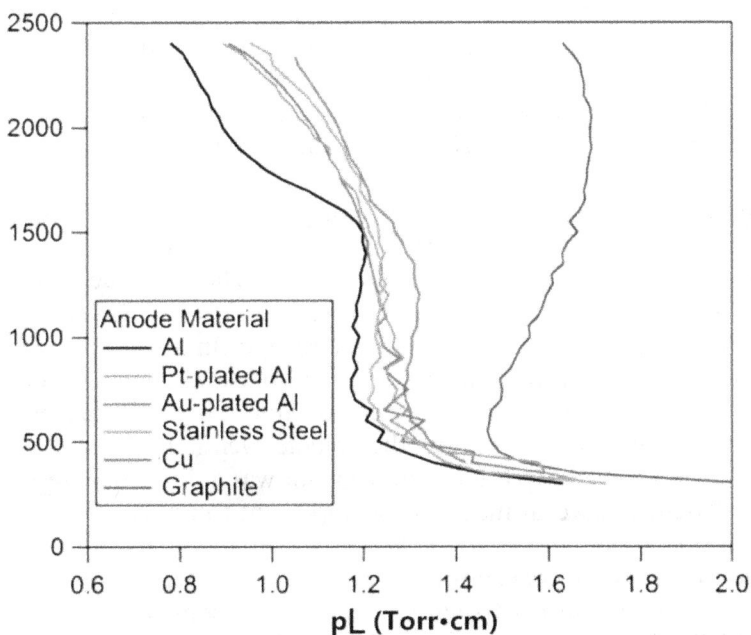

Figure 4.6. Paschen curves (breakdown voltage in V) in helium with copper cathode and different anode materials. From [12].

Figure 4.7. An automated system for the production of reliable Paschen curves over an extended range of pL (left). A wineglass discharge tube. Distance between the cathode and the anode is 2 mm (as shown) (right). From [12].

breakdown up to breakdown value. The first regime was suitable for investigation of single-valued parts of the Paschen curves, while the second regime allowed for studies of the multivalued Paschen curves, and was most suitable for the research conducted in this work. The Paschen curve for any particular anode material was measured 10 times and all curves shown in figure 4.6 were each averaged over 10 curves.

As a result of non-monotonic curves, Penning found that breakdown may occur at three different values of the voltage [14]. Figure 4.8 demonstrate experiments with similar multi-value breakdown points in helium [12]. The pL value was first held at 1.6 Torr cm, while the voltage between the cathode and anode was slowly increased from zero to about 400 V, where breakdown occurred. In a second experiment, for pL of near 1 Torr cm, the voltage between the cathode and anode is increased from zero to 1.5 kV. Then, the pL value was increased without breakdown from 1 Torr cm to 1.6 Torr cm. Afterwards, decreasing the voltage eventually led to breakdown at about 1.2 kV. This somewhat paradoxical behavior was a direct manifestation of the multi-valued Paschen curve. It may also be important to the gas desorption due to the electron heating, etc.

Thus, not only does the concept of $\alpha(E/p)$ lose its physical meaning here but the breakdown at small pL seems to vary with many uncontrollable factors. In the present state of the art, a more or less reliable calculation of breakdown voltages appears to be quite problematic. On the other hand, the calculations for the right

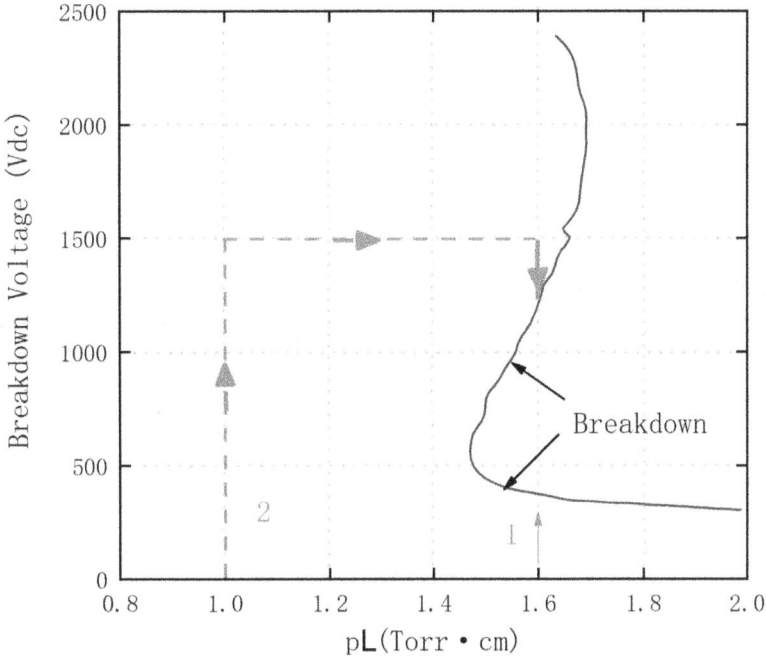

Figure 4.8. Plot of two experiments demonstrating multi-value breakdown points for a graphite anode accessed by (1) increasing and (2) decreasing the applied voltage. From [12].

branch, in which the ionization can be described by the local coefficient α/p as a function of E/p, can be made by using the dependence of γ_{eff} on E/p.

4.4 Cathode boundary conditions for discharge on the right-hand branch of the Paschen curve [15]

The electron diffusion in the cathode region may be considerable, so let us derive the breakdown criterion with its account. We will re-write the breakdown condition equation (3.6), introducing $\lambda_E = T_e/eE$ (for estimations below assuming that $\varepsilon_1 \approx \varepsilon_i$).

$$n_e'' - n_e'/\lambda_E + \alpha n_e /\lambda_E = 0. \tag{4.13}$$

The solution of equation (4.13) for non-zero electron density on the cathode, $n_e(0) = n_{e0}$, (cf equation (3.2)) is

$$n_e(x) = n_{e0} \exp(x/2\lambda_E)\sinh((L - x)\sqrt{1 - 4\alpha\lambda_E}/2\lambda_E) $$
$$/ \sinh(L\sqrt{1 - 4\alpha\lambda_E}/2\lambda_E). \tag{4.14}$$

A local description of the ionization is applicable to the right branch of the Paschen curve

$$\alpha\lambda_E \ll 1. \tag{4.15}$$

4-11

The expansion of equation (4.14) over small parameter (4.15) yields the well-known Townsend relationship (2.2)

$$n_e(x) \approx n_{e0} \exp(\alpha x) - n_{e0} \exp(\alpha L)\exp((x - L)/\lambda_E) \approx n_{e0} \exp(\alpha x). \qquad (4.16)$$

It follows from equations (4.14) and (4.15) that the electron flux even at the cathode

$$j_{e0} = -D_e n_e' + u_e n_{e0} = -u_e n_{e0}\alpha\lambda_E + u_e n_{e0} \approx u_e n_{e0} \qquad (4.17)$$

is defined by the drift. In other words, if condition (4.15) holds, the contribution of electron diffusion is small and its account does not change the relationship (2.2) for the electron current multiplication from the cathode. Therefore, the relationship between the currents j, $j_e(0)$ and the coefficients γ and γ_{eff} is expressed as equations (4.4) and (4.5) within the fluid model. Since the fluid model (and the concept of u_e) is applicable only at a distance $x > \lambda_E$ to the discharge boundary, a further specification of the boundary conditions seems to be of little value in this approximation [15]. With the assumption of $\lambda = $ const and taking into account that $dn_e/dx \approx \alpha n_{e0}$ from equation (4.14) we have

$$n_{e0} - 2n_{emit} \approx -2\lambda\alpha n_{e0} < n_{e0}u_e/\bar{V} \approx 0, \qquad (4.18)$$

i.e. $n_{e0} \approx 2n_{emit}$. The density of emitted electrons n_{emit} in equation (4.18) is related to the current $j_e(0)$ according to equation $j_{emit}(0) = n_{emit}\bar{V}$, where $\bar{V}$ is the average velocity of the emitted electrons. From equation (4.18), we have $j_{e0} = n_{e0}u_e \approx 2j_{emit}(0)u_e/\bar{V}$, i.e. $\gamma_{eff} \approx 2\gamma u_e/\bar{V}$.

To follow the details of the electron drift from the cathode, one should use a kinetic analysis. As an example we discuss now deriving an expression for f_{es} within a one-dimensional model. The kinetic equation for the isotropic EDF component in variable x and total energy $\varepsilon = mV^2/2 - eEx$ for inelastic electron energy balance in the field $eE\lambda > \varepsilon_1\sqrt{2\,m/M}$ is written as

$$\frac{\partial}{\partial x}\left(\frac{w}{3N\sigma(w)}\frac{\partial f_0(\varepsilon, x)}{\partial x}\right) = Nw\sigma^*(w)f_0(\varepsilon, x), \qquad (4.19)$$

where σ and σ^* are the cross sections of elastic and inelastic scattering, varying with the kinetic energy $w = mV^2/2 = \varepsilon + e\varphi(x)$. The electron motion with the initial energy ε is illustrated in figure 4.9. At $\varepsilon < \varepsilon_1$ and $x < (\varepsilon_1 - \varepsilon)/eE$ the kinetic energy of the electrons is less than ε_1 and they move with their total energy ε preserved. Since $\sigma^* = 0$, the differential flux

$$\Phi(\varepsilon) = \frac{V^2}{3}f_1(\varepsilon, x) = -\frac{V^2}{3N\sigma(w)}\frac{\partial f_0(\varepsilon, x)}{\partial x} \qquad (4.20)$$

is also preserved. At $x > (\varepsilon_1 - \varepsilon)/eE$ an electron after an inelastic impact abruptly loses the excitation energy ε_1 (it goes down to the lower step $(\varepsilon - \varepsilon_1)$, as is shown in figure 4.9). Only such electrons are further picked up by the field and cannot return to the cathode. When the two-term expansion of the EDF is valid (when $\sigma \gg \sigma^*$), the characteristic spatial $\lambda_e^* = \sqrt{\lambda\lambda^*}$ and energy $T^* = eE\lambda_e^*$ scales of the EDF variation

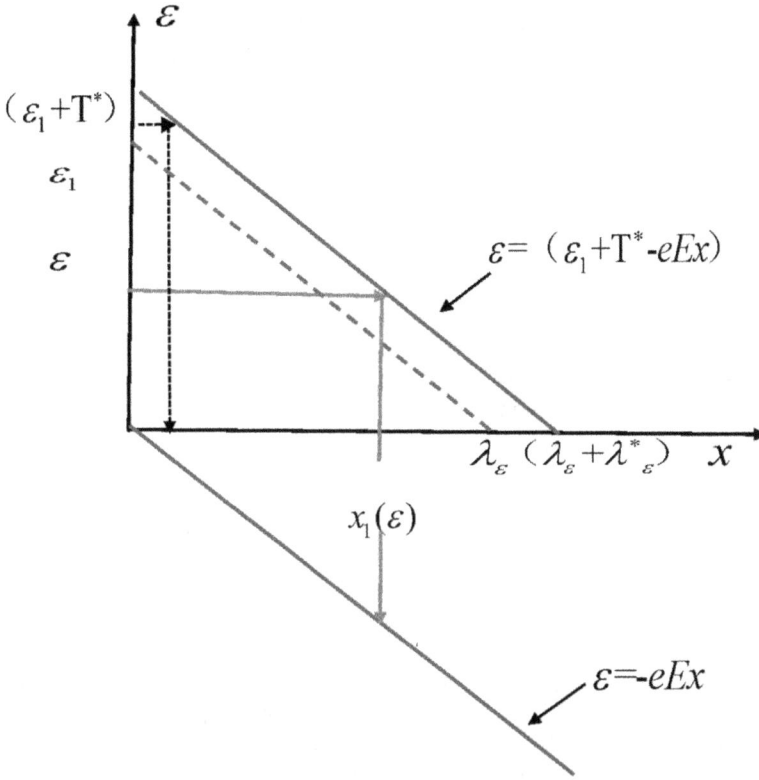

Figure 4.9. Trajectory of an emitted electron in phase space ε, x. After [15].

are small at $\varepsilon > \varepsilon_1$, as compared with $\lambda_E \approx \varepsilon_1/eE$ and ε_1, respectively. Therefore, the EDF varies abruptly above the threshold ε_1 and can be calculated within the black wall model with the zero boundary condition at the energy $(\varepsilon_1 + T^*)$ at a distance $(x_1(\varepsilon) + \lambda_\varepsilon^*)$ (see figure 4.9). The solution to equation (4.19) can be expressed as

$$f_0(\varepsilon, x) \Big/ \int_x^{x_1(\varepsilon)} \frac{\sigma(\varepsilon + eEx)dx}{(\varepsilon + eEx)} = f_0(\varepsilon, 0) \Big/ \int_0^{x_1(\varepsilon)} \frac{\sigma(\varepsilon + eEx)dx}{(\varepsilon + eEx)} = \Phi(\varepsilon), \quad (4.21)$$

with $x_1(\varepsilon) = (\varepsilon_1 + T^* - \varepsilon)/eE = \lambda_E + \lambda_\varepsilon^* - \varepsilon/eE$. As $\lambda_E \gg \lambda_\varepsilon^*$, the contribution of fast electrons with $w > \varepsilon$ to the current and density is small, of the order of about $\lambda_\varepsilon^*/\lambda_E \ll 1$. Marshak's relation [3] for the EDF

$$j_+/\bar{V} = f_0/4 + f_1/2, \quad j_-/\bar{V} = f_0/4 - f_1/2 \quad (4.22)$$

yields the boundary condition on the cathode

$$j_+(\varepsilon, 0) = j_{emit}(\varepsilon, 0), \quad (4.23)$$

where $j_{emit}(\varepsilon, 0)$ is the flux of electrons emitted by the cathode. The substitution of the EDF (4.21) into equation (4.23) gives the mono-energetic electron source on the cathode

$$\frac{1}{f_{es}(\varepsilon)} = \frac{\gamma(\varepsilon)}{\gamma_{eff}(\varepsilon)} = \frac{1}{2} + \frac{\varepsilon}{4} \int_0^{x_1(\varepsilon)} \frac{N\sigma(w)dx}{w(\varepsilon, x)}. \tag{4.24}$$

As the ratio $V^3/\nu = w/N\sigma(w) \approx const$ is approximately fulfilled in Ar, Kr and Xe, we obtain from equation (4.23) that the EDF linear along the coordinate

$$f_0(\varepsilon, x) = f_0(\varepsilon, 0)(1 - eEx/(\varepsilon_1 + T^* - \varepsilon)). \tag{4.25}$$

The expression (4.24) for f_{es} coincides with equation (4.4) with $\bar{V}/(4u_e) = \lambda_E/(2\lambda) \gg 1$. It is easy to derive a general expression for an arbitrary source $f_{emit}(\varepsilon)$ at any $\sigma(w)$ dependence from equations (4.4) and (4.21)

$$\frac{1}{f_{es}} = \frac{\gamma}{\gamma_{eff}} = \frac{1}{2} + \frac{\int_0^{\varepsilon_1+T^*} f_0(\varepsilon, 0)\sqrt{\varepsilon}\,d\varepsilon}{4\int_0^{\varepsilon_1+T^*} \frac{f_0(\varepsilon,0)d\varepsilon}{\sqrt{\varepsilon}\int_0^{x_1(\varepsilon)} \frac{N\sigma(w)dx}{w}}}. \tag{4.26}$$

The emitted electron energy has a more or less uniform distribution in the range from 0 to ε_e, with $\varepsilon_e = \varepsilon_i - 2e\varphi_e$ for the potential emission by ions or $\varepsilon_e = \varepsilon_1 - e\varphi_e$ for the emission by metastable atoms. Here ε_i, ε_1 and $e\varphi_e$ are the ionization energy, the gas excitation energy and the work function of the cathode material [16]. Then we can assume $f_0(\varepsilon, 0)$ in equation (4.26) to be independent of energy as far as $\varepsilon \leqslant \varepsilon_e$, i.e. $f_0(\varepsilon, 0) = 3n_{e0}/2\varepsilon_e^{3/2} = const$. For example, with $V^3/\nu \simeq const$ from equation (4.26), we have equation (4.4) with

$$\lambda_E \approx (\varepsilon_e^2/(6eE\varepsilon_1))\ln((\sqrt{\varepsilon_1} + \sqrt{\varepsilon_e})/(\sqrt{\varepsilon_1} - \sqrt{\varepsilon_e})).$$

In any case, the majority of emitted electrons will return back to the cathode at $\lambda_E \gg \lambda$, and the effective secondary emission yield will be small, in accordance with equation (4.4).

Thus, the secondary emission yield from the cathode, γ_{eff}, in the conventional relation equation (4.7) for the right branch of the Paschen curve, where the local description holds, is not genuine surface characteristic γ. The value of $\gamma_{eff} \ll \gamma$ changes with the field and initial energy of emitted electrons. In the left branch, however, where the field is high and γ_{eff} approaches γ, the local coefficient α has no sense and formulas of the type of equation (4.7) become invalid.

4.5 The time evolution of the breakdown

The criterion (4.7) was derived from the stationarity condition equation (4.6) and describes only a simple electron multiplication, $\mu = 1$. In order to make the minimum 'seed' current go up to a macroscopic value, it is necessary to have, at least, a small 'over-voltage' $\Delta U = U - U_b > 0$, which will provide an increasing electron reproduction, $\mu > 1$. If a single electron leaves the cathode at the initial moment of time, it produces $\mu > 1$ secondary electrons in the second cycle, whose duration is determined by the multiplication time on the cathode; the number of electrons in the third cycle will be μ^2, and so on. The ionization and, hence, the current will exponentially rise in time, as is usually observed in the breakdown. The time of the breakdown evolution after the voltage is applied to the gap is also known as the *delay time*. It has two components: (1) the time for a seed electron to appear, which is the statistical delay time, and (2) the time between the appearance of the first electron and the establishment of a stationary discharge. The statistical delay varies with the intensity and geometry of the primary ionization, the state of the electrode surfaces, etc. If necessary, it can be reduced by using a pre-ionization by an external source. The time of the discharge establishment is defined by the dominant mechanism of cathode emission. If the emission is due to positive ions, the time is defined by the ion drift from the anode to the cathode, $\tau_i = L/u_i$; the electron flight to the anode for the time $L/u_e \ll \tau_i$ can be neglected. Then the ion current density on the cathode at the moment t is

$$j_i(x = 0, t) = \int_0^L \alpha(x) j_e(x, t - x/V_i) dx. \tag{4.27}$$

In order to find the expression for the current rise not only at the breakdown voltage, $(\mu \gtrsim 1)$, but also at the pre-breakdown voltage, $(\mu < 1)$, the boundary condition on the cathode

$$j_e(0, t) = j_0 + \gamma_{\text{eff}} j_i(0, t) \tag{4.28}$$

must preserve the initial cathode current induced by an external source, $j_0 = j_e(x = 0, t = 0)$. Using the relation for the electron multiplication in the field

$$j_e(x, t) = j_e(0, t) \exp\left(\int_0^x \alpha(x) dx\right) = j_e(0, t) M(x) \tag{4.29}$$

where M is the multiplication coefficient in equation (2.3), we get an equation for the ion current density across the cathode:

$$j_i(x = 0, t) = \int_0^L dx \alpha(x) [j_{e0} + \gamma j_i(x = 0, t - x/V_i)] \left[\exp \int_0^x \alpha(E(x')) dx'\right]. \tag{4.30}$$

At $\alpha L \gg 1$, the majority of ions are produced near the anode where the electron current is highest, so we can take $j_i(x = 0, t - \tau_i)$ from under the integral sign in equation (4.30) to find the condition for the discharge maintenance

$$j_i(x = 0, t) \approx [j_0 + \gamma_{\text{eff}} j_i(x = 0, t - \tau_i)]\left[\exp\left(\int_0^L \alpha(E) dx\right) - 1\right]$$
$$= \mu[j_i(x = 0, t - \tau_i) + j_0/\gamma_{\text{eff}}], \tag{4.31}$$

where $\mu = \gamma_{eff}(M - 1)$ is the charge reproduction coefficient equation (4.7). Expressions (4.28–4.31) for the cathode electron current yield [1, 18]

$$j_e(0, t) = j_{e0} + \mu j_e(0, t - \tau_i) \approx j_0 + \mu\left[j_e(0, t) - \tau_i\frac{dj_e(0, t)}{dt}\right]. \tag{4.32}$$

The solution of equation (4.32) with the initial condition $j_e(0,0) = j_{e0}$ is written as

$$\frac{j_e(0, t)}{j_0} = \frac{\mu}{\mu - 1}\exp\left(\frac{(\mu - 1)t}{\mu\tau_i}\right) - \frac{1}{\mu - 1}. \tag{4.33}$$

Clearly, the current rise at $\mu < 1$ is limited by the limiting value of $j_e(0, t \to \infty) = j_0/(1 - \mu)$. This agrees with Townsend's first experiments (chapter 2). At $\mu > 1$, the current increases exponentially with time. It is clear from equation (4.33) that the characteristic time of a breakdown is defined by the value $\tau_i/(\mu - 1)$ and decreases fast with the over-voltage $\Delta U = U - U_b$ due to the exponential dependence of the coefficient a on U.

4.6 Limitations of the avalanche breakdown mechanism

The Townsend breakdown discussed above is not accompanied by noticeable field distortions due to the space charge and in the simplest case can be considered to be uniform over the whole gap. In this, it differs from a streamer breakdown at high pressures, when a thin spark arises between the electrodes, with the adjacent gap regions, which are also in the field, remaining unionized. Then the space charge induced during the evolution of a single avalanche, becomes essential. To conclude, Townsend's theory was considered for a long time to be universally valid for breakdowns at low and high pressures because it provided approximately correct breakdown voltages. The reason for this was a weak (doubly logarithmic) dependence of the voltage equation (4.12) on γ. There are, for example, data indicating that the γ value in the air breakdown varies from 10^{-2} at low pressure to 10^{-8} at atmospheric pressure and large pL, whereas $\ln\gamma$ changes only by 30%, so it was hard to see a deviation from the Paschen law in practice. It was found later, however, that the avalanche theory encountered difficulties in gaps larger than 1 cm at pressures about the atmospheric pressure.

The major difficulty was associated with the time of the breakdown evolution. An avalanche moves at the electron drift velocity. In the avalanche theory, no breakdown may occur before the avalanche bridges the gap at least one time. In reality, everything happens much faster at large L, especially at increasing over-voltage, when the discrepancy between experiment and theory is as large as several orders of magnitude. Under these conditions, a thin ionized streamer travels through the gas between the electrodes, making its way along a positively charged channel; the distortion of the applied field near the streamer is considerable. It has been found that the breakdown criterion equation (4.7) applies only at low voltages and not very large pL values, whereas the streamer mechanism comes into play at $pL > 10$ cm Torr and high over-voltage. The latter case will not be discussed here. Figure 4.10

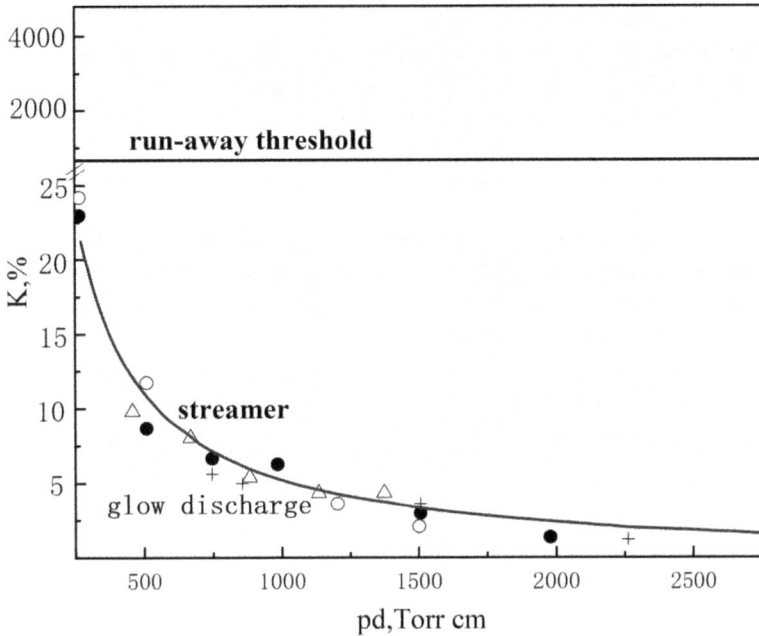

Figure 4.10. Boundaries of realization of different mechanisms of breakdowns. After [17].

illustrates the boundary between the Townsend and streamer breakdowns for helium. One can see that the avalanche mechanism is operative up to high pressures at a low over-voltage.

References

[1] Loeb L B 1939 *Fundamental Processes of Electrical Discharge in Gases* (New York: Wiley)

[2] Thomson J J 1909 Positive electricity *Philos. Mag.* **B18** 821–45

[3] Marshak I S 1961 Electric breakdown of gases at pressures close to atmospheric pressure *Sov. Phys. Uspekhi* **3** 624

[4] Raizer Y P 1991 *Gas Discharge Physics* (Berlin: Springer)

[5] Brown S C 1956 *Handbuch der Physik* vol 22 (Heidelberg: Springer)
Brown S C 1951 High-frequency gas-discharge breakdown *Proc. Inst. Radio Eng.* **39** 1493–501

[6] Francis G 1960 *Ionization Phenomena in Gases* (London: Butterworths Scientific)

[7] Brown S C 1959 *Elementary Processes in Gas Discharge Plasma* (Cambridge, MA: MIT Press)

[8] Von Engel A 1955 *Ionized Gases* (Oxford: Clarendon)

[9] Phelps A V and Petrovic Z L 1999 Cold-cathode discharges and breakdown in argon: surface and gas phase production of secondary electrons *Plasma Sources Sci. Technol.* **8** R21

[10] Golubovskii Y B, Kudryavtsev A A, Nekuchaev V O, Porohova I A and Tsendin L D 2004 *Electron Kinetics in Non-equilibrium Gas-discharge Plasma* (St Petersburg: SPbSU) [in Russian]

[11] Babich L P 2003 *High-energy Phenomena in Electric Discharges in Dense Gases: Theory, Experiment and Natural Phenomena* (Arlington, VA: Futurepast)

[12] Adams S F, Demidov V I, Kudryavtsev A A, Kurlyandskaya I P, Miles J A and Tolson B A 2017 Effect of anode material on the breakdown in low-pressure helium gas *J. Phys.: Conf. Ser.* **927** 012001

[13] Hartmann P, Donko Z, Bano G, Szalai L and Rozsa K 2000 Effect of different elementary processes on the breakdown in low-pressure helium gas *Plasma Sources Sci. Technol.* **9** 183

[14] Penning F M 1931 Anomalous variations of sparking potential as a function of (pd) *Proc. R. Acad. Amst.* **34** 1305

[15] Kudryavtsev A A and Tsendin L D 2002 Cathode boundary conditions for fluid model discharges on the right-hand branch of the Paschen curve *Tech. Phys. Lett.* **28** 621

[16] McDaniel E W 1964 *Collision Phenomena in Ionized Gases* (New York: Wiley)

[17] Starikovskaia S M, Anikin N B, Pancheshnyi S V, Zatsepin D V and Starikovskii A Yu 2001 Pulsed breakdown at high overvoltage: development, propagation and energy branching *Plasma Sources Sci. Technol.* **10** 344–55

[18] Schade R 1937 Über die Aufbauzeit einer Glimmentladung *Z. Physik* **104** 487

Chapter 5

The general structure of a discharge between cold electrodes

In this section the *IV*-traces and axial and structures are examined. Direct or ac current flows in the axial direction and there is no net current to the boundaries. Typically, discharges have a number of specific areas like Cathode fall, negative glow, the Faraday dark space, positive column and anode fall.

5.1 The current–voltage characteristic

When the electrode voltage U exceeds the breakdown voltage, $U = U_b$, a self-sustained discharge is ignited [1]. Formally, this is described by the condition $\mu > 1$ (equation (4.7)), when an electron emitted by the cathode produces more than one electron before drifting away to the anode. The discharge current rises abruptly, on the characteristic time scale $\mu\tau_i/(\mu - 1)$ (see equation (4.33)). An actual circuit always has a resistance R. As the current rises, the voltage drop at R becomes larger, while the electrode voltage decreases. When U drops to U_b, the current stops rising, and a stationary condition is established. It is related to the generator electromotive force (EMF), $\mathcal{E}$, in accordance with the load curve, $\mathcal{E} = U + iR$. By changing R one can change the discharge current in a wide range and observe various self-sustained discharge modes. A typical dependence of the discharge voltage U on the current density j and the parameter pL is presented in figure 5.1 for a stationary discharge in a cylindrical tube between plane electrodes [1]. It is seen that the voltage U coincides with the breakdown voltage at low currents and follows the Paschen curve with changing pL. A low current discharge, in which the effect of the space charge field is negligible, is referred to as a Townsend discharge mode. Let us increase the current by reducing the load resistance R or by increasing the generator *emf*, $\mathcal{E}$.

Figure 5.2 gives the current–voltage characteristics $i(U)$ (*IV*-traces) of the discharge with pL for the right branch of the Paschen curve, $pL > (pL)_m$ (see also figure 5.1). The section BC is for the Townsend mode. The inter-electrode potential

doi:10.1088/978-1-64327-060-9ch5

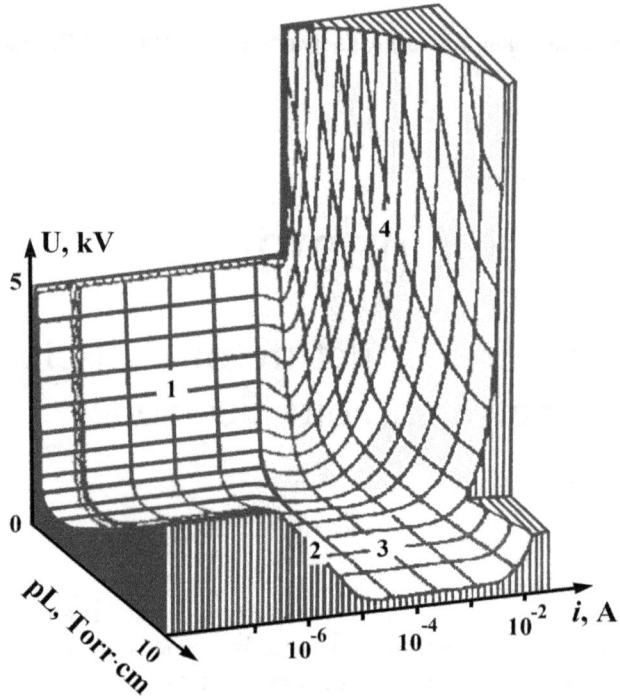

Figure 5.1. Varieties of glows: Townsend (1), undernormal (2), normal (3) and abnormal (4) discharges. After [2].

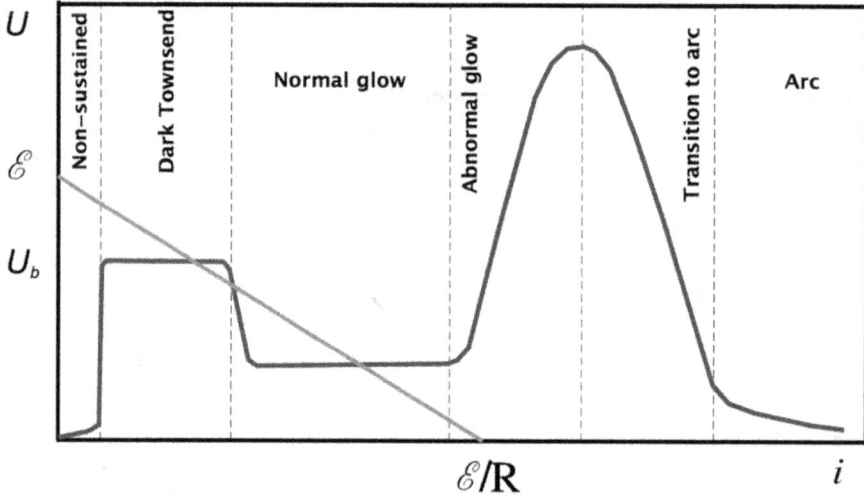

Figure 5.2. *IV*-trace of a discharge between a cathode and anode in a wide range of currents. Non-sustained (AB), dark Townsend (BC), normal glow (DE), anomalous glow (EF), transitional to arc (FG) and arc (GH) discharges. After [2].

remains constant as far as the point C, from which the operation voltage decreases because of the external field distortion by space charges. The CD section is transitional, and its lower part refers to a subnormal mode. Note that this classification of discharge modes has not been generally accepted and the subnormal mode is often combined with the Townsend mode.

As the current rises further, a normal glow is formed, whose voltage is independent of current (like in the Townsend mode) in a wide range of values, sometimes several orders of magnitude (the DE line). In contrast to the Townsend discharge, however, the operation voltage is also practically independent of pL (see figure 5.1). Here only part of the cathode is covered by the discharge, such that the current density does not change with current. Most of the glow volume is occupied by a quasi-neutral plasma. Table 5.1 shows some experimental parameters of a normal glow in various conditions.

When the current becomes so high that all of the cathode area is covered by the discharge, this mode is called an abnormal glow with a rising IV-trace (the EF line). With still further current rise, the cathode heating and its thermal emission become

Table 5.1. Experimental parameters of normal glow discharge with different cathode materials. After [2].

Parameter		Ar	H_2	He	Ne	Hg	N_2	O_2	Air
$(pd)_n$	Al	0.29	0.72	1.32	0.64	0.33	0.31	0.24	0.25
	Cu	—	0.8	—	—	0.60	—	—	0.23
	Fe	0.33	0.9	1.30	0.72	0.34	0.42	0.31	0.52
Torr cm	Mg	—	0.61	1.45	—	—	0.35	0.25	–
	Ni	—	0.9	—	—	0.4	—	—	—
	Pb	—	0.84	–	–	–	–	–	–
	Pt	—	1.0	–	–	–	–	–	–
Un	Al	100	170	140	120	245	180	311	229
	Cu	130	214	177	220	447	208	–	370
	Fe	165	250	150	150	298	215	290	269
V	Mg	119	153	125	94	–	188	310	224
	Ni	131	211	158	140	275	197	–	226
	Pb	124	233	177	172	340	210	–	207
	Pt	131	276	165	152	305	216	364	277
jn/p^2	Al	–	90	–	4	–	–	–	–
	Cu	–	64	–	–	15	–	–	–
	Fe	160	72	2.2	6	8	400	–	–
$\mu A/(Torr\ cm)^2$	Mg	20	–	3.0	5	–	–	–	–
	Ni	160	72	2.2	6	8	400	–	–
	Pb	–	–	–	–	–	–	–	–
	Pt	150	90	5.0	18	–	380	550	–

essential. The glow transforms to an arc (point G), and the cathode current shrinks to a spot with an abrupt voltage drop (the FG line). In the left branch of the Paschen curve (high voltage discharge or obstructed discharge), the Townsend-to-abnormal mode transition usually occurs without a normal glow (figure (5.1)).

5.2 Basic characteristics and spatial structure of the glow

The structure of a glow discharge between two electrodes is quite complex and typically looks like that in figure 5.3. A glow consists of alternating visible dark and light regions [1]. Since all the processes in it are determined by collisions, its characteristic scale is the path length, such that the similarity principle is usually fulfilled: glows with the same value of pL = const differ only in the scale. So it is easier to observe this at low pressures. A narrow Aston dark space (DS) is adjacent to the cathode, followed by (a) thin cathode layer(s) (glow(s)). Then comes a cathode dark space, also known as the Crooks dark space. The gas luminosity in those areas is low while the electric field is higher than in the other discharge regions. Adjacent to this region is a negative glow, which dies down towards the anode to eventually form the Faraday dark space. The latter is followed by a bright positive column. In long cylindrical tubes it is uniform or has a stratified structure which is immobile or

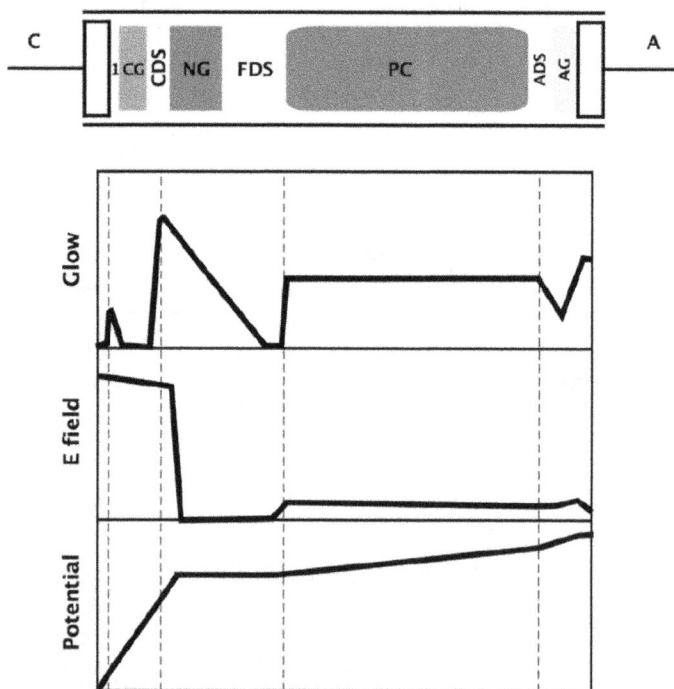

Figure 5.3. The structure of glow discharge (Aston dark space (1), cathode glow (CG), cathode dark space (CDS), negative glow (NG), Faraday dark space (FDS), positive column (PC), anode dark space (ADS) and anode glow (AG)) and distribution of its parameters: luminosity, electric field, electric potential.

moving along the discharge axis striations. Normally, there is a narrow DS near the anode with a thin bright film near the anode surface.

When the position of the electrodes is changed, in particular, when they are brought closer to each other, the positive column becomes shorter, while the potential profiles in the near-electrode regions and the column field E_c remain unchanged. In contrast, the cathode and anode regions move together with the respective electrodes without changing their structure. If the cathode is turned in the tube, all of its parts turn together with it, leaving their positions unchanged relative to the cathode surface.

When the positive column totally disappears, the Faraday space is the first to become shorter, followed then by the negative glow region, although the position of the region boundary on the cathode side does not change. The voltage across the discharge increases, so this mode is called an obstructed, or high voltage discharge. These conditions normally correspond to the left branch of the Paschen curve. When there is no more room for this negative glow edge, the discharge dies.

Figure 5.3 gives the distributions of the basic discharge parameters along the gap length L: the radiation intensity I_r, the potential φ and the field strength E. The high electric field at the cathode falls towards the negative glow boundary. This region is known as the cathode sheath. The field in the anode sheath is much lower. The rest of the discharge volume is occupied by a quasi-neutral plasma, whose field is still lower. The electric field in the positive column is uniform in the absence of strata; its strength in the negative glow region is several orders of magnitude lower than in the sheath and may even reverse its sign there. The field sign in the anode sheath may also change relative to the adjacent plasma (see section 7.6).

A classic device for the glow ignition and study is a discharge tube representing a glass cylinder of radius $R_d = 0.5...5$ cm and length $L = 10...100$ cm with metallic electrodes—a cathode and an anode. For the typical operating gas pressures $p = (10^{-2}...10^2)$ Torr, the characteristic voltages across the electrodes are $U = (100...1000)$ V and currents $I = (10^{-4}...1)$ A. The positive column is not as bright as the negative glow and usually has a different color. For example, the cathode glow in neon is yellow, the negative glow is orange, and the positive column is red. In nitrogen, they are pink, blue and red, respectively. These characteristic spectra of gases are widely used in advertisement tubes. Of special practical interest are the plasma regions: the positive column and the negative glow. Designers of lighting technologies are especially interested in the long positive column, which always occurs in long narrow tubes. To reduce the total voltage drop, a hot cathode is used producing an arc instead of a glow. Other things being equal, the positive column in a glow with cold electrodes and in an arc are identical. Phenomena occurring in a positive column are not affected by the electrode material or state; the column glow is generally invisible in wide tubes or spheres. The size of the electrode sheath varies with the gas and its pressure, the current density, etc. This type of discharge was called a glow because of the nature of the negative region.

References

[1] Raizer Y P 1991 *Gas Discharge Physics* (Berlin: Springer)
[2] Golubovskii Y B, Kudryavtsev A A, Nekuchaev V O, Porohova I A and Tsendin L D 2004 *Electron Kinetics in Non-equilibrium Gas-discharge Plasma* (St Petersburg: SPbSU) [in Russian]
[3] Kudrayvtsev A A, Smirnov A S and Tsendin L D 2010 *Physics of Glow Discharge* (St. Petersburg: Lan) [in Russian]

Chapter 6

The Townsend and subnormal modes

In this chapter the Townsend discharge, in which the effect of the space charge is negligible, will be considered [1–4]. The important parts of the consideration are the conditions for the applicability of the traditional hydrodynamic approximation. For the discharge on the right branch of the Paschen curve it is possible to use the first Townsend coefficient α, which depends on the local value of the electric field and the effective coefficient for ion–electron emission γ_{eff} is substantially reduced due to kinetic effects. On the left branch of the Paschen curve, the situation is more complex. The field here is rather high and the phenomenon of runaway electrons begins to play a significant role. As a result, a kinetic analysis is required. As the discharge current rises, the electric field distortion by the space charge begins to play a role and the discharge becomes unstable. The development of this instability leads to the establishment of a mode of normal current density. In this mode, ionization becomes non-local [5].

6.1 The current–voltage characteristic

We have mentioned in the previous chapter, a Townsend discharge operates at a voltage equal to the ignition voltage U_b. In a one-dimensional geometry, the total current density is constant, $j = j_e + j_i = \text{const}$. Hence, neglecting the diffusion, we have

$$j_e/j = \exp(-\alpha(L - x)), \quad j_i/j = 1 - \exp(-\alpha(L - x)). \tag{6.1}$$

The ion current greatly exceeds the electron current in most of the gap volume. To illustrate, at $\gamma = 0.01$ and $\alpha L = 4.6$, the j_e value reaches j_i only at $x = 0.85L$. Since the electron mobility is much higher than the ion mobility, the difference in the charge densities is still greater. At $b_e/b_i = 100$, the ratio $n_i/n_e = (b_e/b_i)(j_i/j_e)$ is as large as unity only at $x = 0.998L$. This indicates that the ion space charge is dominant in most of the gap volume as shown in figure (6.1). This space charge

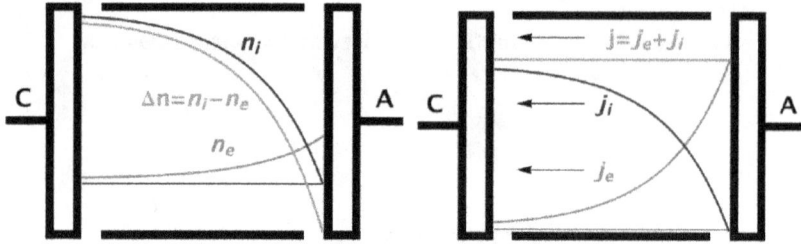

Figure 6.1. Distribution of densities of electron and ion currents (right) and charges (left), for the case when external electric field do not disturbed by the space charge.

may, in turn, distort the external field. Therefore, the condition for the discharge self-maintenance is

$$\mu = \gamma_{\text{eff}}\left(\exp\left[\int_0^L \alpha(E(x)/p)dx\right] - 1\right) = 1. \tag{6.2}$$

Note that a similar expression is valid for a non-plane geometry (see equation (4.9)).

The integration in equation (6.2) is made over the gap length. An avalanche traveling from the cathode to the anode must produce a definite number of electron generations; this number is defined only by the secondary emission yield and is independent of whether the field is uniform or not. Experimental data show that a Townsend glow with $U = U_b$ does take place in the right and left branches of the Paschen curve in a wide range of low currents (see figure 6.1).

Consider the effect of a small (ion) space charge on the operating voltage $U = U_b + \Delta U$. A change in the field changes the coefficient $\alpha(E)$ in equation (6.2):

$$\alpha(E) = \alpha_0 + \alpha'(E_0)\Delta E + \alpha''(E_0)(\Delta E)^2/2, \tag{6.3}$$

with $E_0 = U_b/L$. For example, with assumption (2.4), we will have

$$\alpha' = \alpha Bp/E^2, \quad \alpha'' = \alpha'(Bp/E - 2)/E. \tag{6.4}$$

An estimation from equation (6.1), giving $j_i/j \approx 1 - \exp(-\alpha(L - x))$, shows that one can take the ion density to be constant anywhere except for a small anode region of $\sim 1/\alpha$ (figure 6.1) and the electron density to be low. So the Poisson equation, which in this case has the form

$$d(\Delta E)/dx = 4\pi e n_i, \tag{6.5}$$

gives a field profile linearly falling away from the cathode:

$$\Delta E(x) = \Delta U/L - 4\pi e n_i(x - L/2). \tag{6.6}$$

The field rises relative to U/L in the cathode region but decreases in the anode region. Its distortion becomes greater with rising current, which is transported mostly by ions in nearly the whole gap, $j \simeq eb_i E n_i$. With equation (6.6), the number of multiplications in equation (6.2) is

$$\int_0^L \alpha(E)dz = \alpha_0 L + \alpha' \Delta U + 2\alpha'' L(\pi e n_i L)^2/3 = \ln(1 + 1/\gamma). \qquad (6.7)$$

Since the field perturbation ΔE is assumed to be finite but small, the second-order terms in equation (6.7) were preserved. Because the condition for the breakdown and self-maintenance mode at low currents equation (4.7) is $\alpha_0 L = \ln(1 + 1/\gamma)$, equation (6.7) gives the reduction of the operation voltage relative to the ignition voltage:

$$\Delta U = 2\alpha'' L(\pi e n_i L)^2/(3\alpha'). \qquad (6.8)$$

Therefore, we have a parabolic IV-trace:

$$\Delta U = U_{br} - U = 2\pi^2 \alpha'' j^2 L^3/3\alpha'. \qquad (6.9)$$

It is clear from equation (6.9) that the field distortion by the space charge can either decrease (at $\alpha'' > 0$) or increase (at $\alpha'' < 0$) the discharge operation voltage. A critical point is the inflection in the $\alpha(E/p)$ curve, which, according to the Townsend breakdown condition (4.7), is also the inflection in the ignition $U_b(pL)$ curve. For approximation (2.4), the inflection parameters (6.3) and (4.12) are

$$E_{inf} = Bp/2, \quad (pL)_{inf} = 2.72(pL)_m, \quad U_{inf} = 2.72 U_m/2. \qquad (6.10)$$

To conclude, the field distortion by the space charge at $pL < (pL)_{inf}$ retards the multiplication ($\mu < 1$), leading to a higher operation voltage. The rising IV-trace is stable to perturbations even in the absence of load resistance in the circuit. Since $(pL)_{inf} \sim (pL)_m$, this corresponds to the left branch of the Paschen curve indicating a uniform operation at low gas pressures; this discharge mode is known as a high voltage mode.

It should be stressed, however, that the left-side breakdown is characterized by the electron runaway, for which the coefficient $\alpha(E/p)$ is meaningless, so the considerations above have only a qualitative nature.

The right-hand branch corresponds to a falling IV-trace at $pL > (pL)_{inf}$ because the multiplication $\int_0^L \alpha dz$ (equation (6.7)) grows with current at given U, i.e. the voltage must fall with increasing current. The reason for this is that the integral values in equation (6.2) for a non-uniform field are determined practically only by the high field regions because of the exponential field dependence of α. To satisfy the condition of equation (6.2), the potential drop across the gap must be smaller than in a uniform field in the absence of space charge. If the working point is far off in the right-hand branch of the Paschen curve, the maximum fields are practically the same at different working points.

6.2 Townsend discharge instability in the right branch of the Paschen curve [5]

The IV-trace of a transversely uniform discharge is unstable, since $\alpha'' > 0$. Experimental data show that this instability may lead to a constriction across the current and/or to fluctuations of the discharge current and voltage. The experimental *IV*-trace (figure 5.2) has three sections corresponding to the three discharge modes. The left side is for the Townsend mode, the right one is for the normal mode, and the intermediate section is for the subnormal mode. If the *IV*-trace is measured during the current rise and fall, the Townsend and normal sections coincide with a high accuracy. The transition from the subnormal to the Townsend mode may have a considerable hysteresis, and the constriction may be accompanied by intensive fluctuations [6].

Let us analyse the relationship (6.7) for instability relative to small two-dimensional perturbations

$$j_i(t, y) = j_i^0 + \tilde{j}_i(x) \exp(\Omega_1 t + iky), \tag{6.11}$$

$$\delta_1 E = \Delta E_0 + \delta_1 \tilde{E}(x) \exp(\Omega_1 t + iky), \tag{6.12}$$

where ΔE_0 is defined by equation (6.6) and $n_i = j_i^0/(eb_i E_0)$. For this, we will use equation (4.31), omitting j_{e0}. Bearing in mind that the voltage variation $\delta_1 U$ is independent of the transverse coordinate y, we find from equation (6.5), similarly to equation (6.6),

$$\delta_1 \tilde{E}(x) = 4\pi e \tilde{j}_i(x)(L/2 - x)/(eb_i E_0). \tag{6.13}$$

The number of multiplications can be written as (cf with equation (6.7))

$$\int_0^L \alpha(E)dx = \alpha_0 L + \alpha' \Delta U + \alpha'' \int_0^L (\Delta E_0^2(x)/2 + \Delta E_0(x)\delta_1 \tilde{E}(x))dx. \tag{6.14}$$

Because ΔE_0 is small, we can use the expansion

$$\exp\left(\alpha'' \int_0^L \Delta E_0 \delta_1 \tilde{E} dx\right) \approx 1 + \alpha'' \int_0^L \Delta E_0 \delta_1 \tilde{E} dz, \tag{6.15}$$

and calculate the integral in the right-hand side of equation (6.15), using equations (6.6) and (6.13):

$$\int_0^L \Delta E_0 \delta_1 \tilde{E} dx = \frac{L^3}{12} \frac{(4\pi)^2}{(b_i E_0)^2} j_i^0 \tilde{j}_i. \tag{6.16}$$

Finally, by substituting $\tilde{j}_i(t + \tau_i) - \tilde{j}_i(t) \approx \tau_i(d\tilde{j}_i/dt)$, we find from equation (4.31)

$$\frac{d\tilde{j}_i}{dt} = \alpha'' \frac{L^3}{12} \frac{(4\pi e j_i^0)^2}{\tau_i(b_i E_0)^2} \tilde{j}_i = \Omega_i \tilde{j}_i. \tag{6.17}$$

It is clear from equation (6.17) that the characteristic time for the instability evolution

$$\tau_{sn} = 1/\Omega_i = \frac{12}{(4\pi e n_i)^2 \alpha'' L^3} \tau_i = \frac{3b_i E_0}{4\alpha''(\pi L j_i^0)^2} \tag{6.18}$$

6-4

is defined by the ion drift time. Formula (6.18) can be re-written with the variables pL, j/p^2, and α/p. With the maintenance condition $\alpha L \approx \ln(1 + 1/\gamma)$ and substituting from $\alpha'' = \alpha(Bp)^2/E_0^4$ for the right branch, we find the relation between the increment and the discharge parameters as the scaling law:

$$\tau_{sn} = \frac{3(b_i p)}{4\pi^2} \frac{(E_0/pB)^2}{(\alpha/p)} \frac{(E_0/p)^3}{(j_i^0/p^2)^2} \frac{1}{(pL)^2 p} \approx \frac{3}{4\pi^2} \frac{(U_{br}/pBL)^2}{\ln(1/\gamma + 1)} \frac{(U_{br}/pL)^3}{(j_i^0/p^2)^2} \frac{(b_i p)}{(pL)p}. \quad (6.19)$$

The development of this instability is hampered by the transverse spread of electron avalanches. Since this process is associated with free electron diffusion, the respective large decrement must, at first sight, strongly suppress the instability. It turns out, however, that the effective decrement is not large. For the short time $L/(b_e E)$, the avalanche traveling from the cathode to the anode spreads out at a distance $\Delta y \approx \sqrt{(T_e L)/(eE)}$. This is repeated after the long time $\tau_i = L/V_i$ necessary for the ions produced by the avalanche at the anode to drift back to the cathode. Although the multiplication period is $(\tau_i + L/b_e E)$, the transverse spread occurs only during the small (electronic) fraction of the period. Since the squared displacements are summed up during random walks, the resulting spread for the time $t \gg \tau_i$ is $y^2 \simeq (\Delta y)^2 t/\tau_i = b_i T_e t$. This corresponds to diffusion with the effective coefficient $D_a = b_i T_e$, of the order of the ambipolar coefficient, in spite of the fact that there is, as yet, no plasma in the gap. The respective decrement is $\Omega_d = (D_a k^2)$ [7]. As the increment Ω_i in equation (6.18) does not vary with the wave number, the most undesirable perturbations are those with the minimum decrement (minimum k), which are defined by the cathode radius R_c. The decrement for them is about $\Omega_d^{min} \approx (D_a/R^2)$. When the increment Ω_i is equal to the decrement Ω_d^{min}, the current is such that a transversely non-uniform discharge becomes unstable. A uniform operation of a low current discharge in the right branch is provided only if the breakdown voltage across the gap is turned on for a time shorter than τ_{sn} (equations (6.18) and (6.19)). This normally occurs when the electrodes are made from a special material and have a special design and when a pulsed (sinusoidal) voltage is applied. It is also believed that dielectric electrodes charged by deposited electrons can improve the operation characteristics and maintenance of a uniform high pressure discharge (see, for example, [9]).

To conclude, even a relatively small field non-uniformity induced by the space charge leads to a slowly falling IV-trace (equation (6.9)) and may initiate an operation instability. The farther the working pL value is in the right branch of the Paschen curve, the smaller is the field distortion by the space charge and the lower is the discharge current, at which the instability is initiated. The details of the instability development vary with the circuit parameters, the gap geometry, and the cathode material. Because the instability is due to the exponential function $\alpha(E/p)$, its development stops, roughly speaking, when the maximum field reaches the value (6.10). The non-local ionization in the low field region appears to be essential, so one cannot use the Townsend ionization coefficient $\alpha(E/p)$ defined by the local field. The switching off of the exponential function $\alpha(E/p)$ leads to the establishment of a normal current density j_n, with the discharge occupying only part of the cathode surface. As a result, the glow with an average current density $j < j_n$ splits into regions with the density close to j_n (several orders of magnitude higher than j) and regions of zero-current. In addition to the formation of a stationary discharge with the normal current density, fluctuations of the total current and voltage may also develop. For this to occur, at least one of the conditions is necessary: $R_t < R_{dif}$, $\Omega_i > 1/(R_t \mathscr{C})$, where R_t and $\mathscr{C}$ are the resistance and capacitance of the discharge-circuit system and R_{dif} is the differential resistance of the discharge [7]. Therefore, the lifetime of a uniform subnormal discharge is limited by a transverse instability, which gives rise to a sharply non-uniform potential distribution and, as a result, to the formation of the cathode sheath and normal current density. Note, that it was shown in [10] that a necessary condition for a steady streamer propagation is that the maximum field front should also be of the order of values from equation (6.10). In other words, the state of the streamer tip (leader) should meet the requirement for the transition to a non-local ionization. This circumstance makes us believe that the above instability is directly related to the condition for the avalanche transition to a streamer.

References

[1] Raizer Y P 1991 *Gas Discharge Physics* (Berlin: Springer)
[2] Loeb L B 1939 *Fundamental Processes of Electrical Discharge in Gases* (New York: Wiley)
[3] Von Engel A 1955 *Ionized Gases* (Oxford: Clarendon)
[4] Von Engel A and Steenbeck M 1932 *Elektrische Gasentladungen: Ihre Physik und Technik* (Berlin: Springer)
[5] Kudryavtsev A A and Tsendin L D 2002 Townsend discharge instability on the right-hand branch of the Paschen curve *Tech. Phys. Lett.* **28** 1036
[6] Phelps A V and Petrovic Z L 1999 Cold-cathode discharges and breakdown in argon: surface and gas phase production of secondary electrons *Plasma Sources Sci. Technol.* **8** R21
[7] Kaganovich I D, Fedotov M A and Tsendin L D 1994 Ionization instability of a Townsend discharge *Tech. Phys.* **39** 241
[8] Kudrayvtsev A A, Smirnov A S and Tsendin L D 2010 *Physics of Glow Discharge* (St. Petersburg: Lan) [in Russian]
[9] Nikandrov D S and Tsendin L D 2005 Low-frequency dielectric-barrier discharge in the Townsend mode *Tech. Phys.* **50** 1284
[10] Dyakonov M I and Kachorovskii V Y 1988 Theory of streamer discharge in semiconductors *Sov. Phys. JETP* **67** 1049

Chapter 7

The 'short' (without positive column) glow discharge

As was discussed in chapter 5, the self-organization of a glow discharge is primarily associated with the potential redistribution along the gap length, so that the current transported mainly by ions in the cathode region is later transformed to the electron current. The necessary ionization is initiated by electrons accelerated by the field in this region. To maintain the electron current in the plasma regions, primarily in the positive column, a low field is sufficient for the ionization to compensate for the slow-electron escape to the walls (together with the ions), recombination and attachment. The discharge gap is split into the space charge sheaths and plasma regions with sharp boundaries between them: the boundary width is small as compared with the size of the plasma region or the sheath.

The main specificity of the electrode region is that the electron and ion fluxes in a homogeneous plasma (e.g. in the positive column if the gap is long enough) must be matched with the boundary conditions. Over 99% of the total current in the column is transported by electrons. On the other hand, the electron-to-ion current ratio at the cathode is small, $j_e/j_i = \gamma \ll 1$. Since the enhancement of the partial (electron and ion) flux in any region equals the net ionization in it, there must be a drastic current transformation in the cathode vicinity. The electron-to-ion current transformation requires intense ionization and very high electric fields, many processes characteristic for dc glows occur in the cathode region; the voltage over the cathode fall is quite high, about 100–1000 V, and the luminosity of the negative glow is considerable. The discharge bears its name because of a bright glow.

There is no ion emission at the anode surface and $j_i(L) = 0$. The ion current here varies from zero to its positive column value, which is less than one percent of the total current. So the net ionization rate in the anode region is two or three orders of magnitude less than in the negative glow. As a result, the anode voltage φ_0 is low (< 30 eV), the anode region length is small, and its luminosity is less pronounced than in the negative glow.

doi:10.1088/978-1-64327-060-9ch7   7-1

In a short gap with no positive column in it, the whole discharge essentially represents the cathode region. Since the principal characteristics of a glow discharge are related to its longitudinal structure, we will consider in this chapter a simple case of a short glow between plane-parallel electrodes containing no positive column[1]. The last section of this chapter considers the anode region of the discharge.

The short discharge was chosen because the positive column is not an obligatory part of the discharge: it arises only when the discharge gap is 'long'. In essence, the positive column is a bridge that closes the current when the electrode spacing exceeds some length. It arises in the cold-electrode glow discharge and incandescent-cathode arc discharge [1].

7.1 The fluid model of the cathode region in a normal glow

To predict its behavior in practice, one must have an opportunity to quickly estimate its basic parameters with a simple yet consistent physical model. A simple and still the most popular theory of the cathode voltage fall was developed by von Engel and Steenbeck (see, e.g. [1, 2]). It is based on Townsend's breakdown model with the only difference being that the self-maintenance condition $\mu = 1$ (equation (4.7) with the local $\alpha(E/p)$ dependence (equation (2.4))) does not hold for all of the discharge volume but only for the cathode sheath. It is assumed that the sheath represents an autonomous self-reproducible system, so the gap length L in equation (6.2) must be replaced by the sheath thickness d and the electric field non-uniformity due to the ion space charge must be taken into account. Thus, the condition for the multiplication in the cathode sheath must be written as

$$M - 1 = \exp\left(\int_0^d \alpha(E(x))dx\right) = 1/\gamma_{\text{eff}}. \tag{7.1}$$

Like in a breakdown, a single electron emitted by the cathode induces ionization in the sheath, generating $(M - 1)$ ions. Having returned to the cathode, an ion provides the emission of a new cathode electron. It is assumed that all ions generated by ionization return to the cathode and all of them are utilized for electron production. The dependence of the cathode voltage fall $U_c(pd)$ corresponds to a Paschen curve which has the point of minimum (4.12) and the inflection point (6.10).

We have mentioned in chapter 6 that the transition from a Townsend discharge to a glow may occur in different ways. The field distortion by the space charge on the left of the inflection point at $pL < (pL)_{\text{inf}}$ hampers the electron multiplication, stimulates the IV-trace $U(j)$ rise and stabilizes a high voltage discharge. However, the fluid model for the description of the breakdown and discharge development in the left branch poorly agrees with observations and is invalid because of the electron runaway. On the right of the inflection point, at $(pL) > (pL)_{\text{inf}}$, the current rise produces an instability, because the electron multiplication is facilitated by the

---

[1] An important exception is the contribution of transversal effects to the establishment of the normal current density in the right branch of the Paschen curve.

potential redistribution, such that a transversely non-uniform discharge can operate at a lower voltage. This promotes the normal discharge mode (see chapter 5).

The role of the potential redistribution associated with the plasma production is clear from a qualitative model of the Rogowski glow (1932) [3]. Suppose there is a voltage U higher than the breakdown voltage U_b at a pL value corresponding to the right branch of the Paschen curve. In the absence of the space charge, the potential profile will be linear at first and U/EL. The gain in equation (4.7) is $\mu_1 > 1$, and the respective value of $\alpha = \alpha_1$ will be described by the exponential section of the $\alpha(E/p)$ curve. The charges accumulated at the anode form a plasma region. Since the plasma field is negligible, the field becomes strongly non-uniform: it increases at the cathode and vanishes at the plasma boundary. Because the α values lie along the exponential $\alpha(E/p)$ curve, the gain at constant voltage will be $\mu_2 > \mu_1 > 1$ and the current will rise fast with time (faster than the exponent). As soon as the field reaches a critical value, at which the exponential $\alpha(E/p)$ function is 'turned off', μ begins to decrease. In other words, when the plasma boundary is at a certain distance d to the cathode, μ becomes unity. This is the moment of establishing a stationary discharge with a well-pronounced cathode sheath and plasma region.

In the right branch where the breakdown is due to the local ionization, the discharge development and plasma formation lead to the potential accumulation along a shorter distance (the cathode sheath of thickness $d \ll L$), initiating the transition to a non-local ionization with exponential $\alpha(E/p)$ turned off. This stage is characterized by a fixed normal current density j_n, and the glow covers only part of the cathode at $i < j_n S_c$ (S_c is the cathode area) in such a way that the current density is equal to j_n. This corresponds to the U_{cn} minimum in the $U_c(j)$ curve for the sheath and represents a normal glow with a two-dimensional (2D) current pattern across the cathode. The cathode spot has such an area that the current density in it is $j_n = i/S$, and the potential drop across the cathode is U_{cn}. The discharge voltage U for a partial cathode coverage is independent of the current, exceeding U_{cn} by the value of the potential drop in the positive column. The current rise is accompanied by the growth of the glow spot S at constant j_n and U_{cn}.

Let us now derive a one-dimensional (1D) matching condition for the cathode spot boundary within the local fluid model [4]. If we assume that the instability produces a stationary pattern of the cathode spot with $\sqrt{S} \gg d$, the particle balance equation averaged over the sheath is

$$D_a \frac{d^2 n_i}{dx^2} = n_e Z_i - b_i \frac{U_c\, n_i}{d}, \tag{7.2}$$

where x is the coordinate tangential to the cathode surface, n_i is the average ion density in a thin 1D cathode sheath, U_c is the x-independent sheath voltage, and b_i is the ion mobility. Here we have approximated the ion out-flux from the sheath to the cathode by the right-hand side of equation (7.2). The ambipolar diffusion coefficient describes the transverse spread of the ion profile; the reason for this was discussed in detail in the preceding section. With equation (2.4), the average ionization frequency can be approximated as

$$Z_i = \alpha b_e\left(\frac{U_c}{d}\right) = A p b_e\left(\frac{U_c}{d}\right)\exp\left(-Bpd/U_c\right).$$

As the electron and ion currents in the sheath averaged over d are comparable, we will replace the product $n_e b_e$ in equation (7.2) by $n_i b_i$. Since the sheath thickness d varies with the ion density as

$$d = \sqrt{\frac{U_c}{4\pi n_i e}},$$

the right-hand side of equation (7.2) can be interpreted as a position-dependent 'force'

$$f(n) = -\frac{d\Psi}{dn},$$

with the 'potential energy',

$$\Psi(n) = -\int dn(n_e Z_i - b_i\frac{U_c}{d}\frac{n_e}{d}).$$

So the problem is mathematically equivalent to the problem of heavy body motion over a curved surface. The 'potential' $\Psi(n)$ is convex up at $n = 0$ due to the exponential dependence (2.4) and convex down at a high ion density due to the cancelling of this dependence. So it has two maxima. One is at $n = 0$ corresponding to the zero-current region. In order to describe a smooth transition between two semi-infinite domains, the equivalent heavy body must start its motion at one potential maximum and finish it at the other with zero 'velocity' $\frac{dn}{dx}$. Therefore, it is necessary to have $\Psi(n = 0) = \Psi(n_2)$, with n_2 being the density at the second maximum. This condition yields

$$n_2 = 1.3\frac{B^2 p^2}{e U_c},$$

and the respective field

$$\left(\frac{E}{p}\right)_2 = \left(\frac{U_c}{pd}\right)_2 = B. \tag{7.3}$$

One can see that a stationary boundary in a 1D approximation can exist only between a zero-current domain and a domain (cathode spot) with the normal current density. It follows from this result that the instability can subdivide the discharge into a zero-current region and a cathode spot with normal current density. The electric field (7.3) corresponds to the saturation of the $\alpha/p(E/p)$ curve. This mode exhibits a considerable non-local ionization in the plasma region of the negative glow, requiring a kinetic treatment.

The field profile in the sheath is described by the Poisson equation, in which the field decreases monotonically from the maximum to zero, such that a simple interpolation gives a linear $E(x)$ profile used in the Engel–Steenbeck model of normal glow (see below, equation (7.12)). If the electron density in the sheath is neglected, the Poisson equation yields approximately $4\pi e n_i = dE/dx \sim E_c/d \sim U_c/d^2$. Hence, the current density on the cathode $j = e n_i\, b_i E \sim b i U^2/(4\pi d^3)$ is written as [1]

$$\frac{j_{ns}}{p^2} \simeq \frac{(b_i p) U_{ns}^2}{4\pi (pd)_{ns}^3}. \tag{7.4}$$

In the Steenbeck model, the cathode potential drop U_{ns} and the sheath thickness d_{ns} in a normal glow are taken to be equal to their values at the minimum in the Paschen curve. From equation (4.12), we have [1]

$$(pd)_{ns} = (pL)_m = \frac{2.72}{A} \ln\left(\frac{1}{\gamma} + 1\right);$$

$$(E/p)_{ns} = (E_b/p)_m = B; \tag{7.5}$$

$$U_{ns} = (U_b)_m = \frac{2.72 B}{A} \ln\left(\frac{1}{\gamma} + 1\right).$$

Table 7.1 shows the calculated and measured values of U_{ns}, j_{ns}/p^2, and $(pd)_{ns}$. With the account of some uncertainty in choosing the γ_{eff} value, we get a reasonable agreement between U_n, U_{ns} and j_n/p^2, j_{ns}/p^2. Since the values of $(pd)_{ns}$ and so on in equations (7.5) are calculated from the same formulas as the position of the minimum in the Paschen curve $(pL)_m$ (equation (4.12)), they appear to be close to the experimental $(pL)_m$ values from table 7.1. In contrast, the experimental values of $(pd)_n$ from table 7.1 turn out to be smaller than the $(pL)_m$ values. This is largely because the Engel–Steenbek model neglects the non-local ionization in plasma and the ion return to the sheath from the plasma (see below).

When the current is higher than $j_{ns} S_c$, there is a transition to an anomalous glow accompanied by the voltage rise (figure 5.2). The sheath thickness here is $pd < (pd)_{ns}$ and is described by the left branch of the Paschen curve, so the Engel–Steenbek theory shows a poor fit with the experimental data.

Table 7.1. Measured and calculated parameters of normal discharge. From [15].

Gas	U_{ns}, V		j_{ns}/p^2, mkA(cm Torr)$^{-2}$		$(pd)_{ns}$, cm Torr	
	Calculated	Measured	Calculated	Measured	Calculated	Measured
He	59–177	143	2–5	1.6	1.3–1.45	2.6
Ne	75–220	154	5–18	1.2	0.64–1.62	2.1
Ar	64–165	146	20–160	4.5	0.29–0.33	0.9
Kr	215	196	43	2.1	0.26	1.1
Xe	306	212	16	6.2	0.23	0.7
H_2	94–276	195	64–110	23	0.16–1.0	1.5
N_2	157–233	213	380–400	15	0.31–0.42	0.7

In order to find the transition to a uniform positive column, equations (2.1) of the local model should include the escape of charged particles. Their combined solution with the Poisson equation shows that the field decreases monotonically with distance from the cathode to the column value, whereas the electron density monotonically increases.

7.2 A model of glow discharge with account of non-local ionization

Of all the models in which ionization is assumed to be dependent on the local field the fact is typical that the cathode sheath inevitably transforms into a positive column. In this case, ionization (and glow) concentrates in the cathode sheath, where the field is high. The cathode sheath–plasma boundary coincides with the boundary between the negative glow and positive column, and the Faraday dark space in such models is absent. Such a discharge pattern is in obvious conflict with observations, which attest that ionization in the cathode region is non-local (that is, it does not depend on the local field strength at a given point of the space). Indeed, the electrons emitted from the cathode and also those generated in the cathode sheath and accelerated by the high field in it are responsible for non-local ionization in an adjacent plasma region, where the field is low. Therefore, the negative glow consists of two parts: one occupies part of the cathode sheath, while the other is in the plasma (plasma negative glow, PNG) [5, 6]. The ions generated in the PNG also return to the cathode, providing electron emission from it, and consequently play a major role in sustaining the discharge. This role is especially significant in the case of the abnormal glow discharge, when the cathode sheath is small and ionization in it is weak, so that the overwhelming majority of ions bombarding the cathode originate precisely in the PNG. Therefore, the pattern of the glow discharge is very complicated; specifically, all near-cathode areas of the discharge (including the cathode sheath and negative glow) are autonomous rather than the cathode sheath alone, as it is supposed in the local models.

Thus, the major disadvantage of the available models of the glow discharge is that they use the local field approximation to determine the ionization rate and, as a consequence, assume that the cathode sheath is autonomous. To put it otherwise, they neglect non-local ionization in the PNG. In this situation, the local models will give both quantitatively and qualitatively incorrect results (as was noted in [7]). Therefore, it seems appropriate to develop a more adequate model than the Engel–Steenbeck model and still as illustrative as the former that makes it possible to establish the main functional relationships between the parameters of the low discharge with allowance for non-local ionization in the PNG.

Numerical data concerning the glow discharge (in particular, specific values of the external parameters) are today routinely derived by computer simulation. Advanced computational codes allow one to obtain a discharge parameter spatial distribution that corresponds to a space charge layer in the plasma with direct and reverse electric field. It was found that the distribution thus obtained agrees well with experimental data (see, e.g. [8–11]). However, the whole solution of the self-consistent problem is time-consuming and requires that 'equally exact' elements be used. The fact is that the reliability of a model is controlled by a bottleneck, which is the least accurately known element. Therefore, even if methods applied to some blocks of the code are the most advanced and reliable, if other elements are poorly or insufficiently known, the accuracy of results cannot be improved. For example, the non-monotonic profile of the potential and non-local ionization are frequently displayed with various hybrid schemes in which slow electrons are described in terms of fluid dynamics but their transport and kinetic coefficients are calculated as functions of electron temperature T_e rather than of the local electric field. The self-consistent electric field is found by solving the Poisson equation in this case [8–11].

The profile $T_e(x)$ is found by solving an equation of thermal balance for electrons that takes into account not only volume processes but also spatial transfer due to heat conduction (see, e.g. [10]). Therefore, a shift in spatial coordinate arises between the field and electron concentration and temperature; that is, the profile $T_e(x)$ and, hence, the profile of the impact ionization rate, diffuse by length λ_ε of electron energy relaxation. The result of such a computer simulation is a cathode sheath with a high electron concentration and high field followed by a low-field plasma region, where T_e is still high enough to result in noticeable non-local ionization. The reverse field here arises in a natural way in order to suppress the diffusion of electrons toward the anode. This plasma region is viewed as the PNG and Faraday dark space. In other words, a non-local dependence of the discharge parameters appears on the electric field such that a maximal concentration of the plasma falls into a range with low electron temperatures. Unfortunately, despite a formal qualitative agreement with experimental data, these seemingly plausible results cannot be considered as adequately describing processes in the near-cathode region on a quantitative basis. The fact is that, in such an approach, the electron ensemble is considered as a whole and is characterized by averaged parameters, namely, averaged density n_e, averaged energy (temperature T_e), and averaged directional drift velocity u_e. However, the electron distribution function in the near-cathode region is actually non-local; i.e. different groups of electrons behave differently. Accordingly, they cannot be described by averaged parameters and kinetic analysis is needed (for details, see [5, 6]). While the simulation results qualitatively agree with experimental data in a number of parameters, many critical issues are treated incorrectly in terms of such a fluid-dynamic description. Specifically, current transport in the PNG region and the Faraday dark space are related to the behavior of an 'intermediate' group of electrons with a heavily non-Maxwellian distribution [5, 6]. The parameters of intermediate drifting electrons, which transport the electron current, are almost independent of the parameters of electrons from the basic group, which provide the balance of plasma density n_e and (average) electron temperature T_e over the electron ensemble. In turn, these thermal electrons (which have a near-Maxwellian distribution owing to their high concentration) cannot move in the longitudinal direction and therefore do not participate in current transport and their temperature T_e is controlled by heating due to collisions with intermediate electrons. Self-consistent kinetic analysis of electrons is therefore necessary. Detailed consideration of these issues and related errors is beyond the scope of the present book and will be published elsewhere.

Here a simple model is demonstrated that, allowing for non-local ionization in the PNG, makes it possible to calculate the distribution of the basic longitudinal parameters of the plasma and derive an *IV*-trace of the short (without a positive column) glow discharge. To apply this model, it will suffice to use well-known tabulated data for secondary emission coefficient γ and Townsend ionization coefficient α.

7.2.1 Basic concepts of the model [12]

As was noted above, the basic characteristics of the glow discharge can be described using the example of the short discharge (the simplest case), when the positive column is absent because of a small electrode distance. In this case, 1D analysis will suffice[2].

The definition of the short discharge to a certain extent remains uncertain. It seems reasonable that discharges for which the electrode gap is shorter than that corresponding to an inflection point in the Paschen curve ($L < L_{\text{inf}}$) be referred to as short. Using breakdown condition

$$M = \exp \int_0^L \alpha(E(x))dx = 1/\gamma \qquad (7.6)$$

and approximating the ionization coefficient as equation (2.4), it is possible to get for inflection point parameters [1]

$$(pL)_{\text{inf}} = \frac{2.72^2}{A} \ln\left(\frac{1}{\gamma} + 1\right), \ (E/p)_{\text{inf}} = B/2, \ (U)_{\text{inf}} = (EL))_{\text{inf}}. \qquad (7.7)$$

The fact that length $L < L_{\text{inf}}$ is critical in the sense that it separates two drastically different modes of evolution of the discharge after its initiation counts in favor of our choice.

In fact, the *IV*-trace of the Townsend discharge under the action of the space charge after breakdown (see, e.g. [1]) is given by

$$U = U_{\text{b}} - \text{const} \times (Bp/(2E_{\text{b}}) - 1)j^2. \qquad (7.8)$$

It follows from equation (7.8) that, if breakdown takes place at the left of an inflection point in the Paschen curve ($L < L_{\text{inf}}$), the reduced electric field is $E/p > B/2$. Accordingly, the Townsend discharge has a rising *IV*-trace [1] and no ballast resistance is required to sustain it. As the current increases, the Townsend discharge steadily changes to the glow discharge, which is homogeneous over the cross-section and usually occupies the cathode completely. This property of short discharges is widely exploited in practice, specifically, in plasma display panels, where a high stability of the discharge is necessary. Since a pixel of the plasma display panel is essentially a high-pressure glow discharge ($p = 500-1000$ Torr) in

---

[2] Even if the discharge is long, a glowing positive column appears only in gas gaps confined in the transverse direction. In 'wide' gaps, the transition region between the negative glow and anode is dark; that is, discharge in this case is a variant of the corona discharge.

Ne–Xe or He–Xe mixtures, its small size ($L = 100$–500 μm) gives a desired low value of parameter $pL < 5$ cm Torr [13] (for helium and neon, $((pL)_{\inf} = 7$–8 cm Torr).

At $L > L_{\inf}$ ($E/p < B/2$), conversely, the IV-trace of the Townsend discharge descends. Therefore, as the current grows, the discharge experiences instability (subnormal discharge), causing a step transition to the normal form. Eventually a spot with a normal current density arising on the cathode occupies it only partly. With a further increase in the current, the discharge becomes abnormal with an ascending IV-trace and totally occupies the cathode surface [1, 6].

It should be noted that the discharge is usually considered glow up to a clear-cut minimum point in the Paschen curve. This point corresponds to length L_m which is roughly three times smaller than $L_{\inf}$. It follows from equations (2.4) and (7.7) that $(pL)_m = (pL)_{\inf}/2.72$; that is, $(E/p)_m = 2(E/p)_{\inf} = B$. However, point $(pL)_m$ is not a point of discharge switchover. Therefore, the categorization of short (without a positive column, $L < L_{\inf}$) discharges adopted in this book (as the post-breakdown current grows, the discharge remains stable) seems to be more appropriate.

The IV-trace of the discharge will be derived, as usual, from the Poisson equation for the cathode sheath and discharge self-sustainment condition.

When the Poisson equation is applied to the cathode sheath, the electron density is as a rule neglected and it is written as

$$\frac{dE}{dx} = -4\pi e n_i. \tag{7.9}$$

Density n_i of ions is expressed through their flux, $n_i = \Gamma_i/u_i$.
Solutions to equation (7.9) reported in the literature differ from each other. The reason is that the drift velocity versus field strength dependence is approximated by a linear ($u_i = b_i E$) or square root function ($u_i = k_i\sqrt{E}$). Since the electric field is sufficiently large in most of the cathode sheath, the root-square approximation seems to be more appropriate. Using the relationship $n_i = \Gamma_i/(k_i\sqrt{E})$ and standard linear approximation of the field versus distance dependence [1, 5–7, 14, 15], and current conservation equation we can obtain

$$j_i(x) = j - j_e(x) = j\left(1 - \frac{\gamma}{(1 + \gamma)}\exp(\alpha_0 x)\right). \tag{7.10}$$

Since the high-to-low field transition occurs at the sheath–plasma boundary, the sheath field can be found from the boundary condition $E(d) = 0$. With the account of the field dependence of the ion mobility, $b_i = k_i/\sqrt{E}$, the integration of equation (7.9) [5] yields

$$E^{3/2}(x) = 6\pi j(d - x - \gamma(e^{\alpha_0 d} - e^{\alpha_0 x})/(\alpha_0(1 + \gamma))/k_i. \tag{7.11}$$

The field profile from equation (7.11) is

$$\frac{E^{3/2}(x)}{E^{3/2}(0)} = 1 - \frac{x - \dfrac{\gamma}{\alpha_0(1 + \gamma)}(\exp(\alpha_0 x) - 1)}{d - \dfrac{\gamma}{\alpha_0(1 + \gamma)}(\exp(\alpha_0 d) - 1)}. \tag{7.12}$$

In the limiting case of an abnormal discharge with small values of d and ionization rate in the sheath, $\alpha_0 d \leqslant 1$, equation (7.12) yields the cathode potential $U = 2E_0 d/5$, and $E/E(0) = (1 - x/d)^{2/3}$. This distribution can be approximated by a linear function only at $x \ll d$. The ion generation in the sheath decreases dE/dx in the vicinity of the point $x = d$ and brings the field profile closer to a linear function. For this reason, the whole experimental field distribution in a normal and moderately anomalous glow can be approximated fairly well by a linear function with $U = E(0)d/2$. Figure 7.1 shows the experimental function $E(x)$ and its calculation with equation (7.12) [5]. From equation (7.11) with $U = E(0)d/2$ gives the common relationship (7.4) (see also [1])

$$\frac{j}{p^2} = \frac{(k_i\sqrt{p})U_c^{3/2}}{\sqrt{2}\,\pi(pd)^{5/2}\left(1 - \dfrac{\gamma(\exp(\alpha_0 d) - 1)}{(1 + \gamma)\alpha_0 d}\right)} \approx \frac{(k_i\sqrt{p})U_c^{3/2}}{\sqrt{2}\,\pi(pd)^{5/2}}. \tag{7.13}$$

Since at $\gamma \ll 1$ ion current j_i in the cathode sheath is of the order of total current j, j is substituted for j_i in estimate (7.13). Comparison with self-consistent calculations [5] shows that the resulting error is small.

To find another relationship between parameters j, U and d in (7.13), we make use of the discharge self-sustainment condition. It has been already noted that the cathode sheath consisting of the space charge is not an independent system; therefore, some of the ions come to the cathode from the PNG and relationship (7.6) cannot be used as the discharge self-sustainment condition. A correct discharge self-sustainment condition can be derived from the condition of current density constancy over the discharge length,

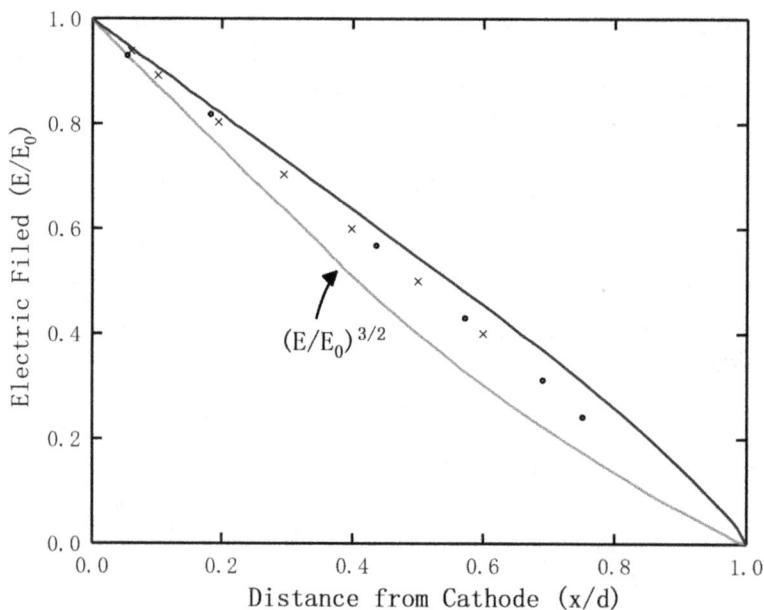

Figure 7.1. Electric field profile in CS: experiment (dots) and calculations (lines). After [5].

$$j(x) = j_e(x) + j_i(x) = j_e(0) + j_i(0) = \text{const.} \tag{7.14}$$

Using equation (7.14) and the boundary condition on the cold cathode,

$$j_e(0) = \gamma j_i, \tag{7.15}$$

we obtain the discharge self-sustainment condition in the form

$$j_e(d)/j_e(0) + j_i(d)/j(0) = 1 + 1/\gamma, \tag{7.16}$$

which is an extension of Engel–Steenbeck condition (7.6).

If the first term alone is left on the left-hand side of (7.15), we arrive at local condition (7.6). However, the second term on the left of (7.15) (the ratio of this term to the first one is sometimes called the plasma efficiency factor [15, 16]) is almost always large and therefore neglect of this term (which means that the cathode sheath is autonomous) is incorrect in the given situation[3].

Electron and ion fluxes appearing in condition (7.15) can be found from the respective balance equations,

$$\nabla \Gamma_{e,\,i} = Z_i x \tag{7.17}$$

where Z_i is an ionization source.

To solve equation (7.16), one must know the ionization parameters in the near-cathode region, where the electric field varies from very high values in the cathode sheath to low values in the PNG. Fast electrons having been accelerated in a high field of the cathode fall are injected into the PNG (where the field is almost absent) with an initial energy far exceeding ionization energy ε_i. These electrons are able to ionize at once irrespective of the local field strength, so that such an electron may travel a large distance before the electron distribution function relaxes to the form corresponding to the local field. In its path, the electron produces much non-local ionization, which by several orders of magnitude exceeds the local ionization, when field E/p in the plasma is weak. In a high electric field of the cathode fall, the situation is aggravated by the effect of electron runaway, when electrons, gaining more and more energy, acquire a non-zero probability of switching to the whistling mode. In this mode, collisions have a minor influence on the motion; that is, the motion becomes ballistic [17].

The aforesaid indicates that use of ionization coefficient (2.4), which depends on local electric field $E(x)$, in the near-cathode region is invalid. Unfortunately, we are unaware of reliable experimental data for the cross-sections of the respective elementary processes (especially for the angular dependences of these cross-sections); therefore, related calculations are inaccurate (exact methods, such as the Monte Carlo method or direct solution of the kinetic equation, are neglected because of their laboriousness).

---

[3] Yet, the local theory gives the *IV*-trace close to the observed one in a wide range. This is because the terms on the left of equation (7.15) are comparable to each other when the current is not too high. Since the first term is strongly (exponentially) dependent on the field, even a small variation of the field in the cathode sheath makes this term prevailing. Formally, this leads to agreement with simplified calculation by (7.6), when the second term on the left of (7.15) is neglected.

To approximately describe ionization characteristics in the near-cathode region of the discharge, we divide the discharge gap into a high-field cathode sheath and a weak-field plasma region, which are separated by a sharp border at point $x = d$. The thickness of the boundary (it is of the order of the Debye screening radius, i.e. much smaller than the thickness of the cathode sheath) will be neglected.

Since ions generated in the cathode sheath under the action of a high field come back to the cathode, it follows from equation (7.16) that the first term in equation (7.15) can be expressed through electron multiplication coefficient M^{CF} in the cathode sheath as

$$j_e(d)/j_e(0) - 1 = M^{CF}(d) = \int_0^d Z_i(x)dx/j_e(0). \qquad (7.18)$$

When finding M^{CF}, one should bear in mind that ionization in the high field of the cathode fall depends on the potential difference overcome by the electron rather than by the local value of the electric field. Since ionization coefficient α is usually tabulated as a function of E/p, the multiplication coefficient in the cathode sheath will be determined by the simplified method [1]. To this end, we replace α by its value in the mean field of the cathode fall, $E = U/d$; that is, it is assumed that $\alpha_{CF} = \alpha(U/d)$. In this approximation, the number of ionizations in the cathode sheath is given by (see [1])

$$M^{CF}(d) = \exp(\alpha_{CF}d) - 1. \qquad (7.19)$$

This approximation corresponds to a constant value of Townsend coefficient α, i.e. to its value in 'effective' field $E = U/d$, which follows from equation (2.4), and is a refinement of the approximation used in [5]. In that work, coefficient α was assumed to be $\alpha_{CF} = \alpha_0 = Ap/2.72$ for any value of the cathode fall, which corresponds to saturation of (2.4) in parameter E/p. Thus, the index of power in equation (7.19) becomes dependent on discharge conditions and can be expressed through coefficients A and B appearing in approximation (2.4),

$$\alpha_{CF}d = Apd \exp(-Bpd/U). \qquad (7.20)$$

Since electrons generated by ionization in the cathode sheath are accelerated in the electric field and also can take part in ionization, the fast electron flux in the cathode sheath grows exponentially. Therefore, the ionization rate also rapidly grows with distance from the cathode and reaches a maximum at the boundary between the cathode sheath and PNG. Broadly speaking, the near-cathode region consists of a cathode dark space, where the fast electron flux is still insignificant, and a glowing region that is part of the cathode glow space[4].

In the PNG, where the field is low, glow and ionization are due to only fast electrons coming from the cathode dark space. Electrons generated here

---

[4] A finer partition of the near-cathode region into the Aston and Hittorf dark spaces, cathode glow space, etc is related to a deeper insight into the dependence of the excitation cross-section on the electron energy; this issue is beyond the scope of this work.

cannot be multiplied and therefore the ionization rate and glow in the PNG can only decrease with increasing distance from the boundary of the sheath. It follows from the aforesaid that the boundary between the cathode sheath and PNG is close to the position of the point where the intensity of source $Z(x)$ is maximal. Since the cross-sections of excitation and ionization behave in a similar way, the ionization and excitation profiles are also close to each other and so a maximum of the discharge is expected to be observed at the cathode sheath–plasma boundary (figure 7.2(a)).

It should be noted that the position of the cathode sheath–plasma boundary remains uncertain. There are many publications where the thickness of the cathode dark space is estimated from visual observations of the discharge glow. Certainly, glow in the immediate vicinity of the cathode is almost absent for two reasons. First, electrons due to ion–electron emission (gamma electrons) are slow, so that excitation and ionization take place at some distance from the cathode. Second, the fast electron flux in the sheath exponentially grows with distance from the cathode. Therefore, the ionization rate and glow intensity also grow in the cathode sheath exponentially and the boundary between the dark and glowing spaces is rather sharp. The thickness of the dark space is frequently identified with the thickness of the cathode sheath. Actually, however, the situation is the reverse: it is the brightest point of the discharge that corresponds to the sheath–plasma boundary. When the fast electron flux responsible for ionization and emission passes into the plasma, it fades out with distance as its slowest electrons decelerate and 'bow out of the game'. For example, in atomic gases, this happens when the electron energy becomes lower than the excitation energy of the first upper level of the atom. It is these 'intermediate electrons' that transport current in the Faraday dark space [5]. Therefore, the negative glow determined visually consists of two parts. This fact should be taken into account in comparing published data. Figure 7.2 shows the longitudinal profiles of the optical and electrical parameters of the discharge.

As has been already noted, the strength of non-local ionization source $Z_i(x)$ in the PNG decays starting from its maximal value at the boundary with the cathode sheath ($x = d$). In the simplest case, it can be approximated, e.g. by an exponential curve first used as early as in [18] and then in [19],

$$Z_i(x) = z_m \exp(-(x - d)/\lambda_{cs}), \quad x \geqslant d. \tag{7.21}$$

In the adopted approximation, the strength of the source at the boundary of the cathode sheath is (see equation (7.19))

$$z_m = \Gamma_e(0)\alpha_{CF} \exp(\alpha_{CF}d). \tag{7.22}$$

In (7.21), λ_{cs} is the characteristic scale of decay, which was tabulated in [19] based on Monte Carlo simulation for argon, helium, nitrogen, and silane. In [20], the values of λ_{cs} for argon are compared with those measured from the spectral line intensity decline in the PNG. It turned out that the experimental values of λ_{cs} for argon exceed

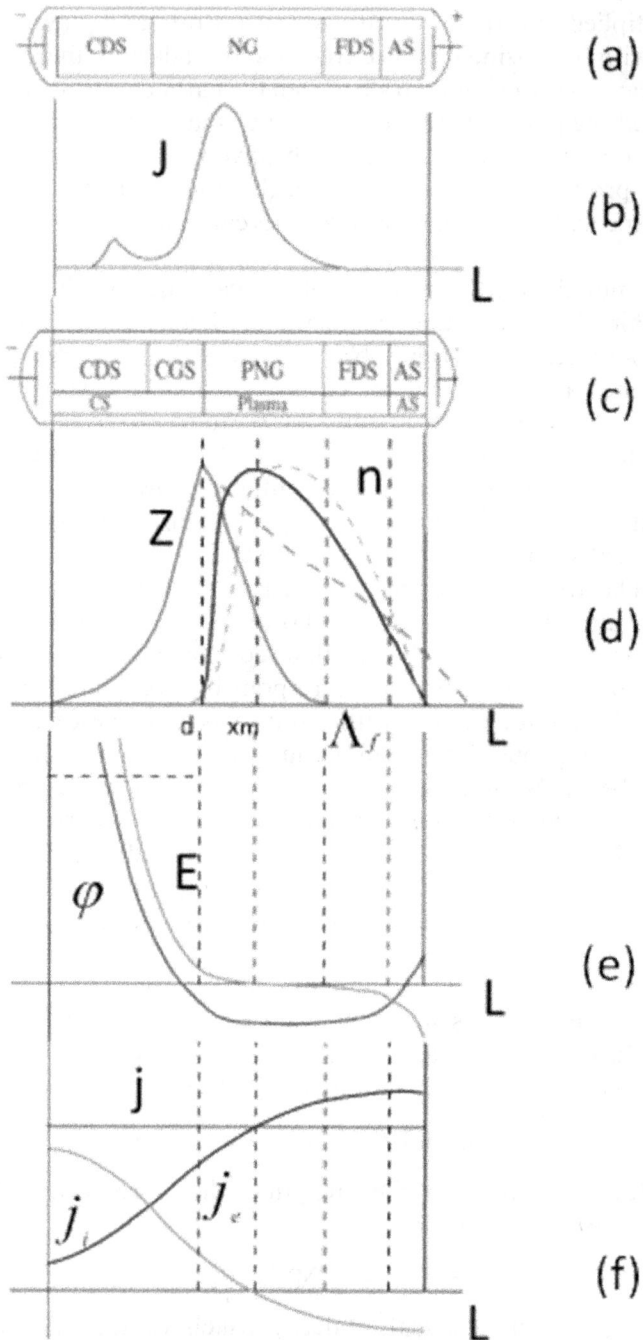

Figure 7.2. Longitudinal distribution of the basic parameters of the short glow discharge (of length L): CDS, cathode dark space; CGS, cathode glow space; PNG, plasma part of the negative glow; NG, negative glow; FDS, Faraday dark space; AS, anode sheath; and Z_i, ionization source. n is the plasma electron density (x_m is the maximum point); Λ_f is the fast electron range; ϕ and E are the electric field potential and strength (dashed line marks the uniform field accepted for the cathode sheath); and j, j_i, and j_e are the densities of the current and its components. From [12].

those calculated in [29] roughly twofold (see [27], figure 8). Later [21, 22], practical estimates for argon were made based on the measured drop of the spectral line intensity rather than on calculated results from [19].

Since reliable data for the scale of λ_{cs} decay are absent, we, in order to be able to rapidly estimate this parameter, relate it to such an important parameter of fast electrons as their range. The range of electrons is defined as length Λ_f which a mono-energetic beam of fast electrons with energy ε travels in a given medium until it completely stops (i.e. until they turn into intermediate electrons).

More or less reliable data for Λ_f are available only for high energies ($\varepsilon > 1$ keV). For the conditions we are interested in (i.e. when values of the cathode fall that represent the upper boundary of the fast electron energy fall into the interval $eU = 200–2000$ eV), available data are sparse. Since even the fastest electrons having gained energy eU stop at the point $x = \Lambda_f$, their range determines the negative glow length in gas discharges [1, 5]. This fact can be used for finding Λ_f under specific conditions.

To estimate $\Lambda_f(U)$, we will make use of the circumstance that, in the energy interval of primary interest for us (from several tens of electron-volts to 1 keV), many dependences of the excitation and ionization cross-sections have a smooth maximum. This maximum is observed in the vicinity of the Stoletov (saturation) point in relationship (2.4) for α and is reached in field $(E/p)_m = B$ (such fields provide conditions for electron runaway [17]). Therefore, Λ_f can be estimated by the empirical formula 7.3

$$\Lambda_f(U) \approx U/E_m = U/(pB). \qquad (7.23)$$

Note that quantity Λ_f, as any integral characteristic, is not sensitive to details of elementary event characteristics; therefore, estimate (7.23) is in good agreement with published data. For argon, the dependence of the range on the applied voltage, which is shown in ([14], figure 5), correlates well with dependence (7.23) with tabulated $B = 180$ V (cm Torr)$^{-1}$. Figure 7.3 plots negative glow length L_{NG} found experimentally versus the cathode fall for various gases ([15], figure 7.6). Also shown are respective dependences (7.23), which are seen to agree well with data from [15].

As was noted, difference $\Lambda_f - d$ is the PNG length, so that the Faraday dark space starts with $x > \Lambda_f$. As parameter pL decreases to a value meeting the position of the minimum in the Paschen curve (when $\Lambda_f > L_m$), the fastest electrons with an energy equal to cathode fall eU reach the anode surface. Since their ionizing capacity is utilized incompletely, the voltage should be raised to sustain the discharge. The discharge in which the Faraday dark space is absent when $pL < (pL)_m$ is called the obstructed discharge [1, 6, 7].

To estimate length λ_{cs} in simple terms, it was suggested that range $\lambda_f(U)$ be identified with the value of difference (xd) at which ionization rate (7.21) becomes low, e.g. falls by $e^2 \approx 10$ times for the sake of definiteness [25]. Then, we get the desired relationship in the form

$$\lambda_{cs} \approx (\Lambda_f - d)/2 \approx U/(2pB) - d/2. \qquad (7.24)$$

$$\Lambda_f, \ cm; l_g, \ cm$$

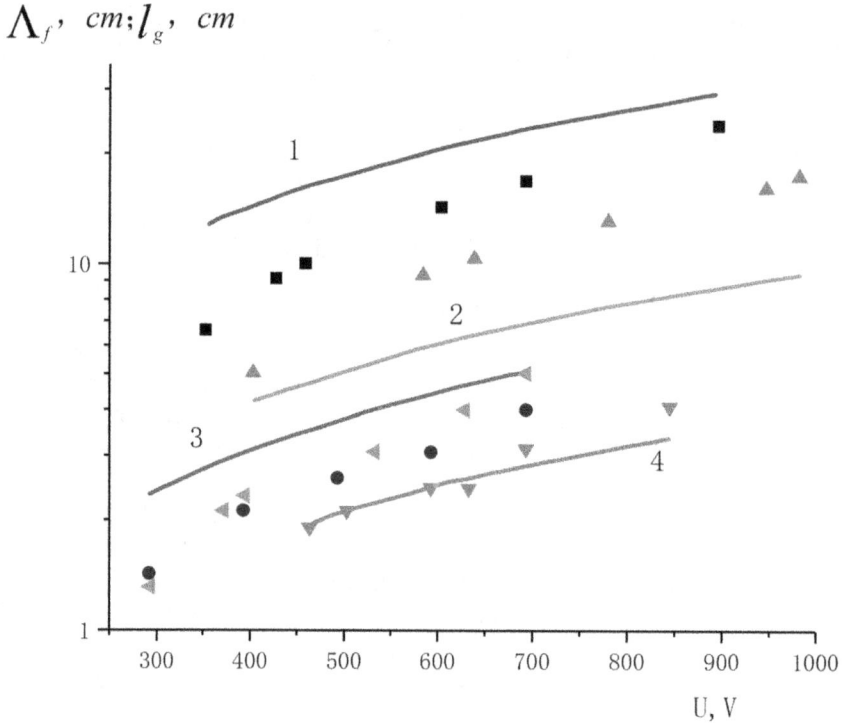

Figure 7.3. Negative glow length l_g found experimentally versus the cathode voltage for (■) helium, (▲) hydrogen, (◄) argon, and (▼) nitrogen [15], and fast electron range Λ_f (7.23) versus the cathode fall for (1) helium, (2) hydrogen, (3) argon, and (4) nitrogen. Data points (●) refer to the fast electron range in argon [14]. From [12].

The simple relationship (7.21) for the source of non-local ionization in the PNG can also be derived by simplifying the model used in [5]. Namely, the field in the cathode sheath is assumed to be uniform or, in other words, the profile of the potential in the sheath is assumed to be linear: $\phi(x) = U(1x/d)$, $x \leqslant d$. Then, using the approximation of fast electron energy continuous losses, we obtain for the effective (allowing for a constant decelerating force) potential in the PNG [5]

$$\phi(x) = U(x - d)/(\Lambda_f - d) \quad x > d. \tag{7.25}$$

Eventually, the model used in [5] yields expression (7.21) for source $Z_i(x)$ where z_m is defined by (7.22) and

$$\lambda_{cs} = (\Lambda_f - d)/(\alpha_{CF}d) = (U - Bpd)/(\alpha_{CF}pd). \tag{7.26}$$

Estimate (7.24) or (7.26) makes it possible to calculate the scale of λ_{cs} decay in (7.21) given voltage U and thickness d of the cathode sheath[5].

---

[5] It should be noted that, if $\alpha_{CF}d \geqslant 2$ and thickness d of the sheath constitutes a tangible fraction of Λ_f, the scale of λ_{cs} decay in (7.26) becomes sensitive to d and Λ_f. In this case, expression (7.26) for λ_{cs} is preferred. At the same time, estimate (7.24) is preferred for the abnormal discharge, when the cathode sheath is thin.

Thus, expression (7.21) combined with (7.24) or (7.26) can be used to find the spatial distribution of the strength of source $Z_i(x)$ and, accordingly, the multiplication coefficient if the PNG,

$$M^{PNG}(x) = \int_d^x Z_i(x)dx/\Gamma_e(0) = e^{\alpha_{CF}d}\left(1 + \alpha_{CF}\lambda_{cs}\left(1 - e^{\frac{x-d}{\lambda_{cs}}}\right)\right). \tag{7.27}$$

According to (7.27), the total number of multiplications (ionizations) in the PNG,

$$M^{PNG}(x) = \int_d^x Z_i(x)dx/\Gamma_e(0) = e^{\alpha_{CF}d}(1 + \alpha_{CF}\lambda_{cs}) \tag{7.28}$$

at $\alpha_{CF} > 1$ exceeds that in the cathode sheath (see (7.19)).

When determining the fraction of ions coming back to the cathode (the second term in discharge self-sustainment condition (7.16)), one should bear in mind the following important circumstance. Unlike in the cathode sheath, where a strong field sends ions back to the cathode, the field in the quasi-neutral plasma that corresponds to ambipolar diffusion is low. Therefore, some of the ions generated in the PNG may move toward the cathode and side walls or recombine in the volume instead of returning to the cathode. For an ordinary plasma with carrier concentration $n = n_e = n_i$ and a carrier mobility independent of the field and concentration, equation of balance (7.17) takes the form (for details, see [1, 5–7])

$$\nabla D_a \nabla n + Z_i(x) - \beta n^2 = 0, \tag{7.29}$$

where $D_a = D_i(1 + T_e/T_0)$ is the ambipolar diffusion coefficient and β is the volume recombination coefficient.

Since the ion density in the layers is small compared with the ion concentration in the PNG, we can impose the zero conditions at the boundary of the cathode sheath and on the anode in (7.29): $n(d) = n(L) = 0$. Then it follows from (7.29) that the profile $n(x)$ has a maximum n_m at the point x_m (see figure 7.1). Accordingly, the ions generated at $x > x_m$ move toward the anode, while those generated at $x < x_m$ return to the cathode. To put it differently, the field changes sign at the point of plasma concentration maximum [1, 5–7, 23]. It suppresses the diffusion of electrons toward the anode (figure 7.2), forms a potential well for electrons (the electrons confined in the well do not contribute to the current at all), and provides the constancy of the current over the gap. The electron current in the Faraday dark space is transported only as the diffusion current of unconfined (intermediate in terms adopted in [5]) electrons. Therefore, a potential jump in the anode sheath is always negative for short discharges ($L < L_{inf}$); that is, it decreases the current of intermediate electrons toward the anode. For the Maxwellian electron distribution function, potential difference U_a between the anode and point of concentration maximum can be roughly estimated by equating the chaotic electron current toward the anode to the electron current (note that this problem demands a kinetic consideration, since the distribution of intermediate electrons may differ strongly from the Maxwellian one),

$$e\frac{n_m}{4}\sqrt{\frac{8T_e}{\pi m}}e^{\frac{eU_a}{T_e}} = j - eD_a\frac{n_m}{L - x_m}. \tag{7.30}$$

The proximity of the point of field reversal to the point of concentration maximum was observed experimentally in [24] and predicted in [8–10]. In [5, 23, 25], this condition was used to find point x_m. At the point of field reversal ($E = 0$), both the electron and ion currents have only diffusion components. Since $D_e \gg D_i$, the total current at point x_m almost coincides with the electron current: $j(x_m) \approx j_e(x_m)$ [23]. If volume recombination plays a minor role and ions move toward the side walls and anode, it follows from equation of balance (7.29) that the second term in (7.16) is

$$j_i(d)/j_e(0) = M^{PNG}(x_m). \tag{7.31}$$

Ultimately, discharge self-sustainment condition (7.16) takes the form

$$M^{CF} + M^{PNG}(x_m) = 1/\gamma. \tag{7.32}$$

One can make use of expressions (7.19), (7.20), and (7.27) to find M^{CF} and $M^{PNG}(x)$ and solve ambipolar diffusion equation (7.29) to find x_m. Being interested in the longitudinal (along the x-axis) distribution of the parameters, we simplify (7.29), representing the ion motion in the transverse direction through characteristic time τ_r of ambipolar diffusion, $\tau_r = (R/2.4)^2/D_a$. Then, upon substituting $Z(x)$ in the form of (7.21), equation (7.29) takes the form

$$\frac{d}{dx} D_a \frac{dn}{dx} + z_m \exp(-(x - d)/\lambda_{cs}) - n/\tau_r - \beta n^2 = 0. \tag{7.33}$$

Since most electrons in the PNG are confined, the electric field does not warm them. If it is taken into consideration that the rate of heat conduction by electrons, which smooths out the electron temperature profile, is high, the ambipolar diffusion coefficient in (7.29) and (7.33) can be regarded as constant. Eventually, equation (7.33) will turn into an inhomogeneous non-linear equation with constant coefficients. If volume recombination is insignificant, equation (7.33) becomes linear and its solution can be expressed in quadratures. If, for example, $R \gg L - d$, we have

$$\frac{n_e(x)}{n_\lambda} = 1 - e^{\frac{d-x}{\lambda_{cs}}} - \frac{x - d}{L - d}\left(1 - e^{\frac{d-L}{\lambda_{cs}}}\right). \tag{7.34}$$

In this case, point x_m where the plasma concentration reaches a maximum is found from the expression

$$x_m = d - \lambda_{cs} \ln\left(\frac{\lambda_{cs}}{L - d}\left(1 - \exp\left(\frac{d - L}{\lambda_{cs}}\right)\right)\right), \tag{7.35}$$

where $n_\lambda = z_m \lambda_{cs}^2/D_a$. Since the profile of source $Z(x)$ coincides with that used in [23], expressions (7.34) and (7.35) are similar to the expressions derived in [23].

Since $n_e \propto 1/D_a \propto 1/T_e$ in (7.34), an uncertainty in the electron temperature directly influences the error in determining n_e. Therefore, the electron temperature should be known exactly to correctly find the plasma concentration. In [10], it was postulated that $T_e = 1 \text{eV}$; in [11], T_e was set equal to 0.1 eV. All other things being the same, the respective values of n_e will differ by one order of magnitude. It is remembered once again in this connection that the accuracy of calculation with any code is governed by a bottleneck, the least accurately known element.

It follows from (7.35) that the position of point x_m in the 1D geometry depends on a single parameter, $\lambda_{sc}/(Ld)$ [23], and is independent of T_e. Given d, L and U_c, x_m is easy to find from (7.35) using expression (7.24) or (7.26) for the scale of λ_{cs}. Substituting (7.35) into (7.27) yields an expression for the effective multiplication coefficient in the PNG, $M^{PNG}(x_m)$, that does not require direct calculation of x_m. Then, from (7.34) and (7.35), we come to

$$M^{PNG}(x_m) = \alpha_{CF}\lambda_{cs}e^{\alpha_{CF}}\left(1 - \frac{\lambda_{cs}}{L - d}\left(1 - e^{\frac{d-L}{\lambda_{cs}}}\right)\right). \tag{7.36}$$

Upon substituting (7.19) and (7.36) into (7.32), discharge self-sustainment condition (7.16) takes a final form (see (7.6)),

$$e^{\alpha_{CF}d}\left(1 + \alpha_{CF}\lambda_{cs}\left(1 - \frac{\lambda_{cs}}{L - d}\left(1 - e^{\frac{d-L}{\lambda_{cs}}}\right)\right)\right). \tag{7.37}$$

The second term in (7.37), which describes ions coming from the negative glow space, is always large; that is, the local models, in which only the first term on the left of (7.37) is present, fail in adequately describing the situation and result in uncontrollable errors. Condition (7.37) combined with relationship (7.13) following from the Poisson equation for the cathode sheath makes it possible to construct an IV-trace of the glow discharge for given gap length L and pressure p.

Thus, the longitudinal distribution of the basic parameters of the short glow discharge has the form depicted in figure 7.1. Figures 7.1(a) and (b) show the main regions of the discharge that are observed visually (see, e.g. [1, 7]). Figure 7.1(c) shows the layers into which the discharge was partitioned. In figures 7.1(b) and 7.1(d), the continuous lines refer to the case when range Λ_f of fast electrons is shorter than the electrode gap ($\Lambda_f < L$) and the dashed lines refer to the case of the hindered discharge, when $\Lambda_f > L$ and some of fast electrons reach the anode, so that their ionizing capacity is utilized incompletely.

7.2.2 Main results from the model [12]

The most complete body of experimental data on glow discharge parameters is that gained for argon. The main results were summarized in review [14], and recent experimental data and simulation results are given in works [20–22], upon which we will rely.

First, it should be noted that, while for the ionization coefficient in argon many experimental data and adequate approximations such as (2.4) are available (see, e.g. [1, 7, 14, 15]), secondary emission coefficient γ remains somewhat uncertain. This is because the electron emission from the cathode in the real situation is caused not only by ions but also by metastable and fast atoms and also by radiation. In addition, emission depends on uncontrollable factors, such as the cathode surface condition, etc. Since these factors are impossible to discriminate in practice, the gas discharge is usually described with the effective coefficient of ion–electron emission γ. This coefficient defined as a ratio of the electron flux entering into the discharge to the ion flux bombarding the cathode is a complicated function of parameter E/p, cathode surface condition, etc (for details, see, e.g. [26]). Today, finding γ from experimental data for breakdown in gases (i.e. from relationship (7.6) with U_b and pL known) seems to be the most consistent and reliable approach. For argon, the most comprehensive data obtained with such an approach were reported in [26], where E/p dependences of γ were derived by processing a large array of experimental data ([26]; figures 5, 11). These results seem to be the most reliable[6].

Figure 7.4 plots γ from [26] versus parameter E/p and the respective approximation in the E/p interval of interest.

Figures 7.5–7.8 compare experimental IV-trace curves from [20–22] (symbols) with curves calculated by the model suggested in this work (continuous lines) and Engel–Steenbeck model (dashed lines).

In all the cases, the voltage equals the breakdown value at low currents (Townsend discharge). As the current grows, the experiment agrees only with the calculation by our model and the local model yields heavily overestimated values of the voltage.

Let us indicate characteristic points of the 1D short ($R \gg L - d$) glow discharge by considering limit cases in greater detail.

For the obstructed discharge ($L < L_m$, $U_b > BpL$), it follows from discharge self-sustainment condition (7.37) that the voltage is a monotonic function of cathode sheath thickness d and, hence, current. Accordingly, if the current density given by (7.13) is such that d becomes smaller than electrode spacing L, the IV-trace starts increasing monotonically with current. It has been already noted that range Λ_f exceeds L in this case; that is, fast electrons with an energy equal to the cathode fall reach the cathode and the Faraday dark space is absent. Under these conditions, we almost invariably have $\lambda_{cs} > Ld$ (except for the case of high currents, when cathode sheath thickness d is much smaller than L). Then, expanding the exponential in (7.37) into a series, we get the self-sustainment condition in the form

$$e^{\alpha_{CF}d}\left(1 + \frac{\alpha_{CF}(L-D)}{2}\left(1 - \frac{L-d}{3\lambda_{cs}}\right)\right) \approx e^{\alpha_{CF}d}\left(1 + \frac{\alpha_{CF}(L-D)}{2}\right) = 1 + \frac{1}{\gamma}, \quad (7.38)$$

which was derived in ([5], formula (52)).

Expression (7.34) then yields a near-parabolic plasma concentration profile,

---

[6] Note that later [27] data reported in [26] were subjected to much correction. The fact is that the glow discharge model with γ taken from [26] gave underestimated results and predicted a drop of the voltage with increasing current, which physically makes no sense. The use of γ as an adjustable parameter [27] radically changed its value compared with that obtained in [26] (see also [11, 21, 22]). If we concede that the approach adopted in [27] is valid, this means that the results obtained in [26] by processing a large data array from many authors become worthless. In our opinion, the discrepancy between the results obtained in [26, 27] indicate the inadequacy of the fluid-dynamic model used in [27] to describe the behavior of electrons to finding the basic parameters of the glow discharge rather than the inapplicability of the data for γ gathered in [26]. Therefore, we used the values of γ from [26] when comparing with experimental data.

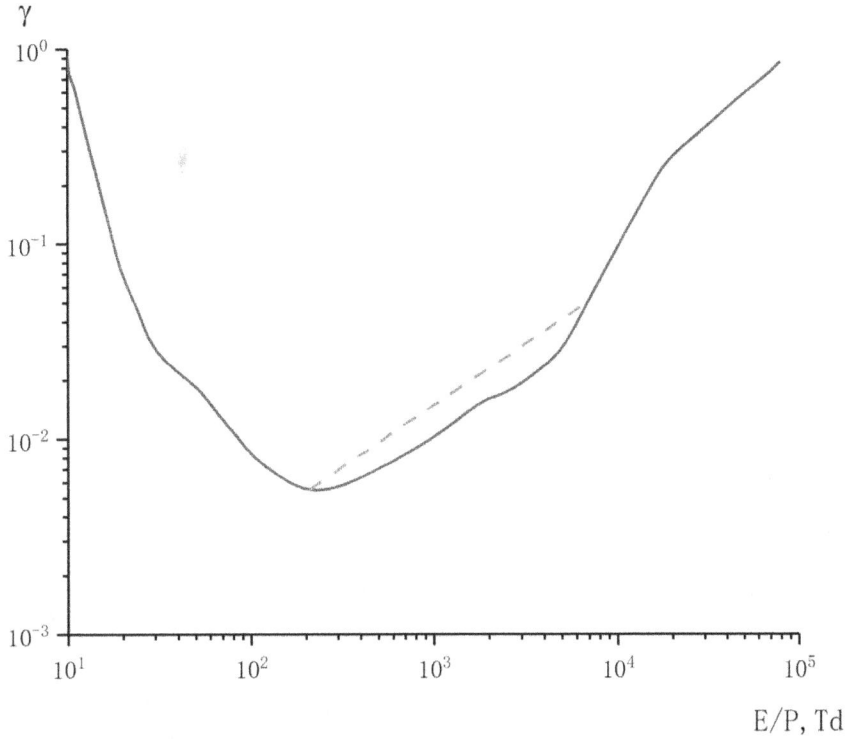

Figure 7.4. Effective coefficient of ion–electron emission in argon. The continuous line, data reported in [26]; dashed line, approximation of the continuous line in the form $\gamma = 0.000\,22(U/(pd))^{0.6}$, where U is given in volts and pd, in cm Torr. From [12].

$$n_e(x) \approx \frac{s_m}{2D_a}(x - d)(L - x), \tag{7.39}$$

in which point x_m of concentration maximum (see (7.35)) is in the middle of the plasma region of the discharge [5],

$$x_m \approx (L - d)/2. \tag{7.40}$$

Under such conditions, half the ions generated in the plasma return to the cathode and the other half go toward the anode.

It is seen in figures 7.4–7.7 that, as parameter pL increases and most ions come back from the plasma toward the cathode, switchover of the Townsend discharge to the glow discharge is accompanied by a small dip in the IV-trace. This follows from discharge self-sustainment conditions (7.37) and (7.38) for the situation with $L_m < L < L_{inf}$, when $BpL/2 < U_b < BpL$. In this case, discharge self-sustainment conditions (7.38) and (7.39) yield a non-monotonic dependence of voltage U on d (and hence on the discharge current). At low currents, when the sheath is thick ($d \approx L$) and $\lambda_{cs} > Ld$, formulas (7.38)–(7.40) are valid. However, as the current rises and d decreases subject to $L > L_m$, a rapid transition to another limit case, $\lambda_{cs} < L - d$, takes place. Now the asymptote of self-sustainment condition (7.37) is given by

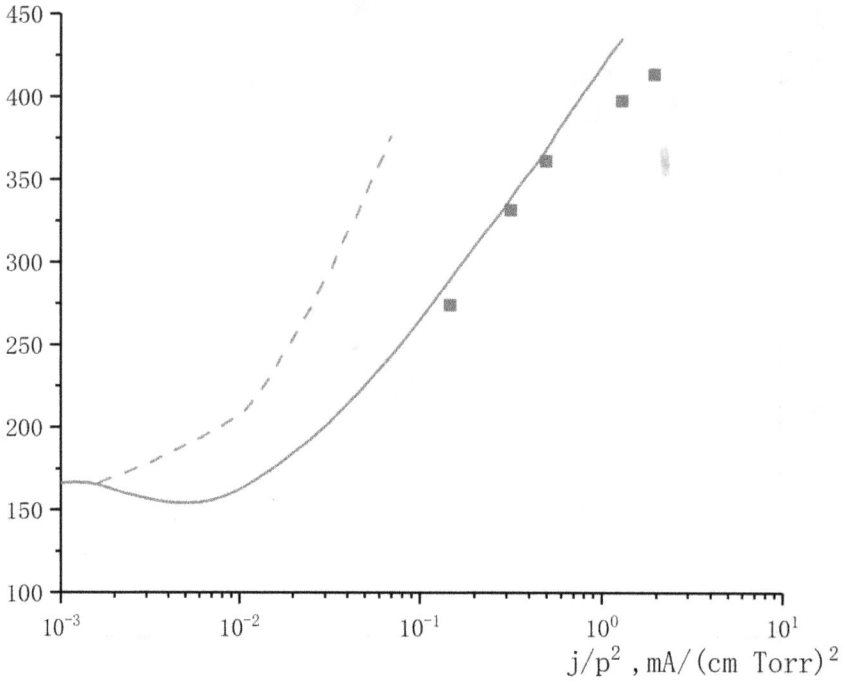

Figure 7.5. *IV*-trace in argon at $pL = 45$ Pa cm. Symbols, data points from [21]; the continuous line, calculation by the model suggested; and the dashed line, calculation by the Engel–Steenbeck model. From [12].

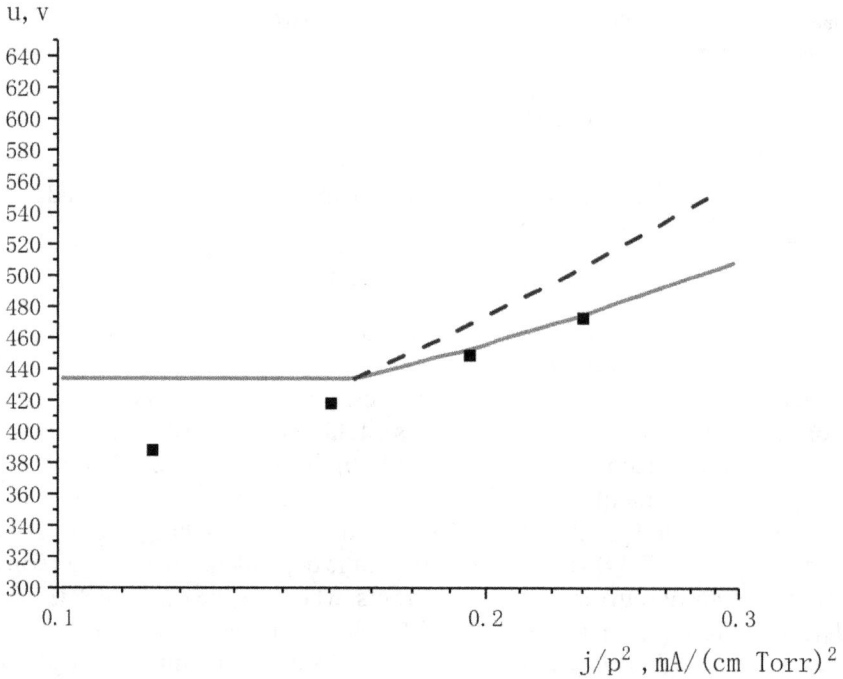

Figure 7.6. The same as in figure 7.5 for $pL = 75$ Pa cm. From [12].

U, V

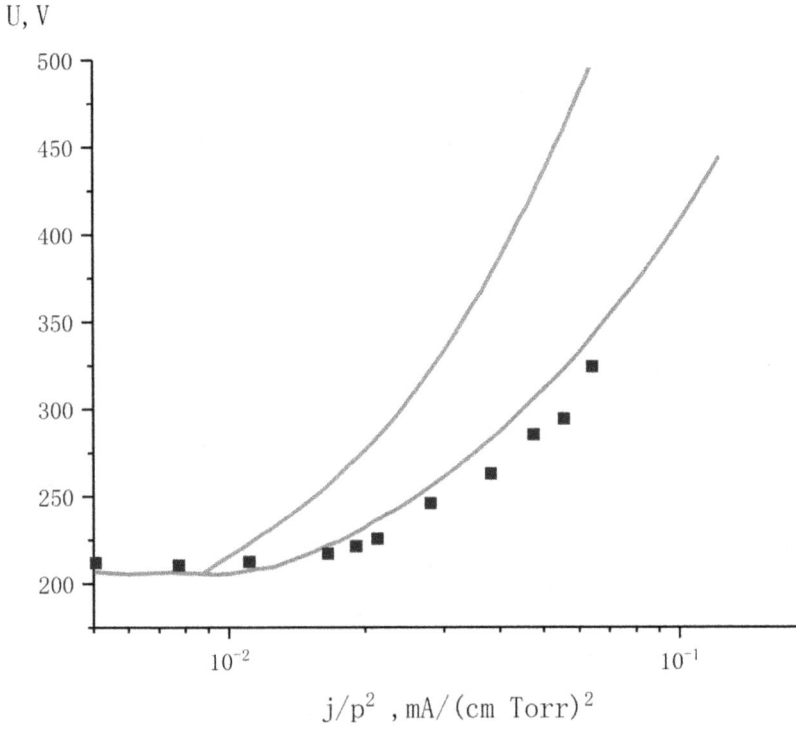

Figure 7.7. The same as in figure 7.5 for $pL = 150$ Pa cm. From [12].

U, V

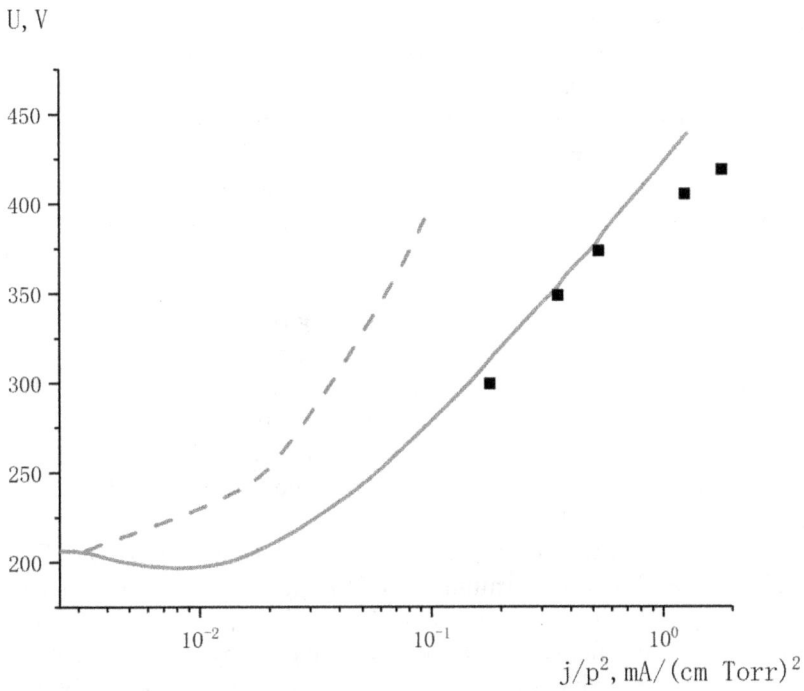

Figure 7.8. The same as in figure 7.5 for $pL = 133$ Pa cm [20]. From [12].

$$e^{\alpha_{CF}d}\left(1 + \alpha_{CF}\lambda_{cs}\left(1 - \frac{\lambda_{cs}}{L - d}\right)\right) \approx e^{\alpha_{CF}d}(1 + \alpha_{CF}\lambda_{cs}) \approx 1 + \frac{1}{\gamma}. \qquad (7.41)$$

In this case, most ions from the PNG return to the cathode, $M^{PNG}(x_m) \approx M^{PNG}$ (see (7.27)). Accordingly, for the plasma concentration profile in the PNG we have from (7.34)

$$\frac{n_e(x)}{n_\lambda} = \frac{L - x}{L - d} - e^{\frac{d-x}{\lambda_{cs}}}. \qquad (7.42)$$

Then, the position of the maximum point (see (7.35)),

$$x_m = d - \lambda_{cs} \ln \frac{\lambda_{cs}}{L - d} \approx d + \lambda_{cs}, \qquad (7.43)$$

is shifted from the boundary of the sheath by λ_{cs}, where λ_{cs} is the characteristic scale of decay of the ionization source strength in the PNG.

In the highly abnormal mode, when $\alpha_{CF}d \ll 1$, self-sustainment condition (7.41) in view of (7.24) transforms into

$$\alpha_{CF}\Lambda_f \approx \frac{U}{2pB}\alpha_{CF}\left(\frac{U}{d}\right) = \frac{1}{\gamma}. \qquad (7.44)$$

Since $U > pBd$ in this case, quantity α_{CF} is close to saturation and varies weakly. If coefficient γ remains constant, the discharge voltage is independent of the current, as follows from (7.44). As far as we know, this phenomenon has not been documented for plane-cathode discharges, while for hollow-cathode discharges, a discharge self-sustainment condition in the form of (7.44), $U/\varepsilon_0 = 1/\gamma$, was derived by Kagan with colleagues [28] as early as in 1976 ($\varepsilon_0 = E/\alpha$ is the 'cost' of generation of an electron–ion pair). Note also that, according to (7.44), discharge voltage U strongly depends on coefficient γ: it varies roughly in inverse proportion to γ. The results of glow discharge simulation shown in ([11], figure 1), where voltage U varies noticeably in response to a slight variation of $\gamma \ll 1$ seem to be consistent with our conclusion. Such a situation cannot be explained in terms of the local model, which predicts only a slight γ dependence of U, $U \sim (1 + \gamma) \approx 1$.

Thus, based on the above analysis, the following procedure for rapidly estimating and predicting the basic properties of the short glow discharge can be suggested.

(1) The *IV*-trace of the discharge can be constructed by solving the Poisson equation for the sheath (or using estimate (7.13)) and using self-sustainment condition (7.37) (or its asymptote (7.38) or (7.41)). The multiplication coefficient in the sheath is calculated by (7.19) and (7.20); in the PNG, by (7.27) with regard to the scale of decay given by (7.24) or (7.26).

(2) The plasma concentration profile in the PNG is found by (7.34) or its asymptote (7.39) or (7.42). Determination of the plasma absolute concentration in the case of the short discharge is not a trivial task. The generally accepted estimates from known current and field ($j = en_eu_e(E)$) carried out by analogy with the positive column here are inapplicable and so related processes should be analysed in a way

similar to that described above. Maximal plasma concentration n_m can be estimated from the ion current toward the cathode, the ion current being a sum of the ion currents arising in the PNG and cathode fall,

$$\frac{j}{1+\gamma} \approx \frac{eD_a n_m}{x_m - d} + j\left(1 - \frac{\gamma}{1+\gamma}e^{\alpha_{CF}d}\right); \tag{7.45}$$

hence,

$$n_m = j\frac{\gamma(x_m - d)}{eD_a(1+\gamma)}(e^{\alpha_{CF}s} - 1). \tag{7.46}$$

It should also be remembered that, to find ambipolar diffusion coefficient D_a and, hence, the plasma absolute concentration, one must know the temperature of slow electrons confined within the potential well. The absolute concentration of the plasma can be found by solving the corresponding balance equation (see, e.g. [5]).

It has already been mentioned that the method suggested in this work requires knowledge of only the coefficient of electron–ion emission γ and parameters A and B used in the approximation of ionization coefficient α. For most gases, they are well known. This greatly simplifies calculations and makes it possible to rapidly estimate discharge parameters for many gases and their mixtures.

Our results are easy to extend for high-pressure discharges and discharges confined in the transverse direction. Since ions subjected to a high field of the cathode sheath return to the cathode, a solution for the cathode sheath in these discharges is similar to (7.13). When analysing processes in the PNG, one should use a solution to the ambipolar diffusion equation in the form of (7.29) or (7.33) with allowance for diffusion toward walls and/or volume recombination losses, which is not a particular problem.

As the discharge current and/or pressure increase, the charged particle escape from the NG and FDS is determined by the *bulk* recombination. At $l_r < \{R_d, (L - x_m)\}$ in equation (7.33), we have a parabolic density fall in the Faraday space $(x > x_m)$

$$\sqrt{\frac{n_m}{n(x)}} - 1 = \frac{(x - x_m)}{l_r}, \tag{7.47}$$

where l_r is the characteristic recombination path.

The estimations show that as long as the gap length is

$$L - x_m < l_r^{2/3}\{[2b_e(x_m - d)/b_i]^{1/3} - 1\}, \tag{7.48}$$

the discharge is short, exhibits a field reversal at $x = x_m$ and the existence of a potential well for electrons. When the gap becomes longer than in equation (7.48) and the inequality

$$eDdn/dx > j \tag{7.49}$$

becomes opposite, there is no field reversal or electron potential well. The Faraday dark space then smoothly transforms to a positive column.

Thus, the autonomous system in a glow discharge involves the whole near-cathode region including the cathode sheath, the negative glow region and the Faraday dark space, rather than the sheath only. For this reason, the discharge *IV*-trace can be found only by a combined solution of the Poisson equation for the sheath which takes into account the ions coming back to the cathode from the negative glow plasma. Even in a normal discharge, the sheath thickness d is smaller than both the negative plasma length, $(x_m - d)$, and $(pL)_m$ at the Paschen curve minimum, decreasing with the current rise in an abnormal discharge. In a short gap with no positive column, the non-local ionization (the presence of fast electrons in the negative glow plasma) leads to the field reversal, creating a potential well for the electrons. The field reversal point corresponds to the maximum density of charged particles and is at $x_m \approx \min\{R_d, x_m\}$[7].

The *IV*-trace of the long discharge ($L > L_{inf}$, $U_b < BpL/2$) has a deeper minimum. Here, the characteristic descends even at the stage of the Townsend discharge (see (7.8)), so that switchover to the glow discharge looks like a sort of instability [6, 29] that eventually results in a normal density of the current [1, 7, 15]. In a long discharge confined in the transverse direction, a positive column may arise. In this case, the field reverses at two points [5]. The first one is the aforementioned point x_m of plasma concentration maximum in the PNG, which produces a potential wall for electrons. The second point of field reversal, x_a, after which the direct field recovers to the value in the column, is at the boundary between the Faraday dark space and the region adjacent to the positive column. In analysis of the longitudinal structure of the near-cathode region, the second point of field reversal can be viewed as a limiting point of this region (virtual anode). In other words, length x_a should be used as an effective length of the discharge instead of L in this case.

7.2.3 The second field reversal and the transition to a positive column

It follows from the above discussion that the formation of an autonomous positive column, in which the ionization is balanced by the electron escape at every point x, should be expected at $\min\{R_d, l_r\} < (L - x_m)$.

When the wall escape is dominant, $R_d < \{l_r, (L - x_m)\}$ and $R_d < (\Lambda_f - d)$, the problem of calculating the profiles from the ambipolar diffusion equation [1] becomes 2D. The assumption that $I(x) \approx$ const for finding x_m becomes invalid. It is hard to derive a simple formula for the position of the field reversal point x_m, so one has to solve a 2D equation for a particular situation. One can see directly from equation (7.33) that with the enhanced inequality $R_d \ll \Lambda_f$, the position of this point tends to a smaller scale, $d + R_d$, rather than to Λ_f like in equation (7.33). With distance from the cathode, $x_m < x < \Lambda_f$, the plasma density profile in the negative

---

[7] The plasma region in a normal discharge at $L \gg \sqrt{S}$ is essentially 2D, expanding towards the anode. Its shape and the electron current distribution across the anode, which are both related to this region, can be found from equation (7.33).

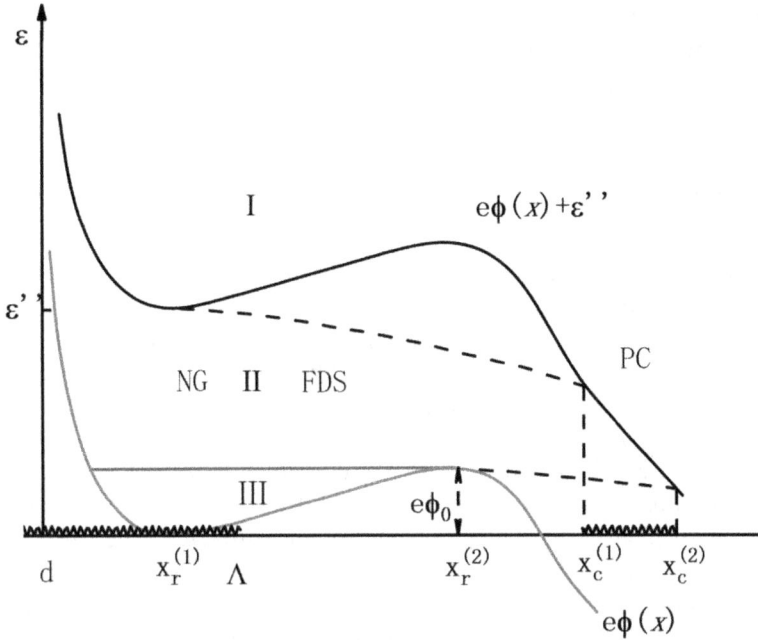

Figure 7.9. Near-cathode axial plasma potential profiles at low pressure [5]: area I corresponds to fast electrons, area II corresponds to fast electrons and area III corresponds to imprisoned electrons. From [5].

glow follows the ionization source profile, $n(x) \approx I(x)\tau_{d\perp}$[8]. At $x > \Lambda_f$, equation (7.33) yields an exponential density fall towards the anode in the negative glow and Faraday dark space

$$n(x) \cong n_m \exp[-2.4(x - x_m)/R_d]. \tag{7.50}$$

The fall has a scale of order of $R_d/2.4$. As a result, the electron diffusive flux sharply reduces with x. So the condition (7.49) cannot be satisfied over the whole discharge length down to the anode, in contrast to the short 1D discharge discussed above. The maintenance of a current-carrying plasma must also involve ionization, and the field must again be directed to the anode. It corresponds to the transition to a positive column. So there are two field reversal points. One point is at the density maximum x_m and is responsible for the formation of a potential well for electrons. The Faraday space length can be easily estimated as a point, at which the plasma density (7.50) decreases to $\sim n_{pc}$ (figure 7.9). This gives an estimate for the Faraday space length at $R_d < \{l_r, (L - x_m)\}$

$$L_F \approx \frac{R_d}{2.4} \ln \frac{n_m}{n_{pc}} \tag{7.51}$$

---

[8] Therefore, the basic assumption of the Engel–Steenbeck model that only the cathode sheath is autonomous becomes valid for narrow tubes at low currents and $R_d \ll d$. A quantitative treatment requires the consideration of 2D effects.

which is about the tube radius. The other field reversal point is at $x = \tilde{x}$, followed by the reconstruction of the dc field in the column. By substituting the distribution (7.50) into equation (7.49) with the discharge current in the column, $j = en_{pc}b_e E_{pc}$, we get an estimate for $\tilde{x}$

$$\tilde{x} = x_m + \frac{R_d}{2.4} \ln\left(\frac{2.4 T_e n_m}{eE_{pc} R_d n_{pc}}\right) = x_m + \frac{R_d}{2.4} \ln\left(\frac{2.4 \lambda_\varepsilon}{R_d} \frac{n_m}{n_{pc}}\right) = x_m + L_F. \qquad (7.52)$$

When the electron energy relaxation path is $\lambda_\varepsilon > R_d$, the transition to a positive column requires a kinetic treatment.

The current transported through the sheath by fast electrons can be described using only the parameters of these electrons. The description of the plasma regions (the plasma part of the negative glow and the Faraday dark space) requires the knowledge of the EDF of slow (plasma) electrons with energy $\varepsilon < \varepsilon_i$. At a distance $d < x < x_m$ in the negative glow plasma, the current of fast electrons is transformed to that of slow intermediate electrons (figure 7.9), whose density is much higher. Their energy relaxation occurs only by quasi-elastic impact, so their relaxation path, λ_ε is large, $\lambda/\sqrt{\delta} \sim 100 \lambda$, and often exceeds the Faraday space length. In this case, the distribution function of the plasma electrons is non-local and varies with the full energy $\varepsilon = w + e\varphi(x)$. Because the behavior of electrons with a given value of ε is unaffected by the other electrons in the absence of electron–electron collisions, it is convenient to subdivide the EDFs of plasma electrons with $\varepsilon < \varepsilon_1$, ε_i into two subgroups: $f_0 = f_{0b} + f_{0s}$. The intermediate electrons with f_{0s} have energies in the range $e\varphi(\tilde{x}) < \varepsilon < \varepsilon_1$. They can overcome the potential barrier at the point of the second field reversal. The most numerous are trapped electrons with $\varepsilon < e\varphi(\tilde{x})$, which do not transport current and do not exhibit the Joule heating. So they have a maxwellian distribution with the temperature T_e due to a strong electron–electron interaction. They are trapped in the potential well and can move only within its limits. Their density obeys the Boltzmann law:

$$n_b(x) = n_m \exp(e\varphi(x)/T_e) \qquad (7.53)$$

and equals zero at $x > \tilde{x}$. The escape of slow electrons is associated with the slow processes of ambipolar diffusion or volume recombination (equation (7.33)). Therefore, the plasma electron densities in the plasma part of the negative glow and Faraday space are largely due to the Maxwell–Boltzmann trapped electrons, which provide the plasma quasi-neutrality in the negative glow. Their temperature T_e is spatially uniform and low, slightly higher than room temperature. It is defined by the integral energy balance of the trapped electrons: the heating in collisions with intermediates having the characteristic average energy $\sim\varepsilon_1$ and the cooling in collisions with neutrals. To calculate the temperature of trapped electrons, one must know the distribution functions of fast electrons with ε, $w \gg \varepsilon_1$, ε_i, producing excitation and ionization, and of the intermediate electrons f_{0s}. Therefore, this is a kinetic problem. The T_e values are necessary for the calculation of the potential profile and the rates of ambipolar diffusion, recombination and stepwise excitation, as well as some other parameters.

In a 1D approximation, $L \gg R_d$, all intermediate electrons diffuse towards the anode with the conservation of the total energy ε; their EDF is described as

$$f_{0s}(\varepsilon, x) \sim \int_x^{x_1(\varepsilon)} \frac{dx'}{D_e(\varepsilon, x')} \int_d^{x'} I(\varepsilon, x'')dx'' \qquad (7.54)$$

with $D_e = V^2\lambda/3$; the upper limit of the integration of $x_2(\varepsilon)$ in the black wall approximation can be derived from the expression $e\varphi(x_2(\varepsilon)) = \varepsilon_1 - \varepsilon$. Since nearly all the electrons produced in the negative glow plasma belong to this group, the density of intermediate electrons, n_s, and their current j_s can be estimated from equation (7.54) as

$$n_s \sim I(x)\tau_s, \quad j_s \sim D_e(\varepsilon_s)dn_s/dx, \qquad (7.55)$$

where $\tau_s \approx L_F^2/D_e$ is the time of the free longitudinal diffusion. Within the approximation

$$dj_e/dx = \alpha_0 j_f, \qquad (7.56)$$

we have

$$j_s(x) \approx \alpha_0 \int_0^x j_f(x')dx' + [j_f(d) - j_f(x_0(x))]. \qquad (7.57)$$

At $x > \Lambda_f$, the intermediate electron current is constant

$$j_s^0 \sim \alpha_0 \int_0^{L_f} j_f(x')dx' + j_f(d).$$

It is clear that the EDFs of the three groups of electrons are associated with different physical mechanisms, so the $T_e(x)$ profiles calculated in terms of the fluid model from the integral energy balance of all electrons (fast, trapped and intermediate) may yield exotic values of the ionization rates, potential profiles, particle densities and fluxes, which would have nothing in common with the actual reality. For the same reason, it seems inconvenient to subdivide the electron current into drift and diffusive currents, as is normally done within the fluid model. The potential difference φ_t between the field reversal points x_m and $\tilde{x}$ is about several T_e values. Since T_e for trapped electrons is rather low, φ_t is small as compared with the first excitation potential, although the density difference may be large (see equation (7.50)). Besides, the reverse field between the points x_m and $\tilde{x}$ is so low that even its mere existence has been questioned until recently. Because the electron current is transported through the Faraday space by intermediate electrons only, it cannot, in principle, be expressed via the total electron density and its derivative. All attempts to introduce thermal diffusion and non-uniform electron temperature (average energy) do not seem to have much perspective either. The ionization by fast electrons is valid as far as the point $x = \Lambda_f$, whereas the ionization by intermediate electrons come into play only at $x > \tilde{x}$ (figure 7.9), at the positive column boundary.

The potential profile in the region between the Faraday space and the negative glow can be estimated from the equations for ions (7.33), omitting the ion diffusion

$$b_i \frac{d(nE)}{dx} = I(x) - \frac{n}{\tau_{d\perp}}. \tag{7.58}$$

At $x < \tilde{x}$, the relation between the plasma density and the potential has the form:

$$n(x) = n_s(x) + n_m[\exp(-e\varphi(x)/T_e) - \exp(-\varphi_t/T_e)]. \tag{7.59}$$

The density of trapped electrons is zero at $x > \tilde{x}$, so the plasma density in equation (7.58) in this region is determined by the intermediate electrons, $n_s(x)$, and must be found with the EDF (7.54). There is no ionization by fast electrons in the transition region, and the source $l(x)$ in equation (7.58) is provided only by intermediate electrons. It is clear from figure 7.9 that ionization is involved only at $x > \bar{x} > \tilde{x}$. Therefore, a more precise Faraday space length than (7.51) is $(\bar{x} - x_m)$. Equation (7.58) yields a parabolic potential profile near the point $\tilde{x}$:

$$e\varphi(x) = e\varphi_t - \frac{e(x - \tilde{x})^2}{2b_i\tau_{d\perp}}, \tag{7.60}$$

where the characteristic time of the side wall escape, $\tau_{d\perp} = R_d^2/[(2.4)^2 D_a]$, includes the ambipolar diffusion coefficient defined by the average energy of intermediate electrons. It is $\sim\varepsilon_1/2$, and the time $\tau_{d\perp}$ is much shorter (by a factor of $\sim\varepsilon_1/T_e \ll 1$) than the respective time in the major Faraday space volume defined by T_e. Therefore, the length of the transition region, $[\tilde{x}, \bar{x}]$, is smaller than that of the Faraday space.

The precise pattern of the transition between the Faraday space and the positive column varies with the specific ionization and recombination processes. In the ionization model, a simple mechanism of standing striation becomes possible when the generation of charged particles correlates with the excitation rate, like in stepwise ionization or when the ionization threshold is close to the excitation threshold. If the Faraday space length L_F is comparable with the energy relaxation path $\lambda_e = \lambda/\sqrt{\delta}$ of intermediate electrons, their energies in the vicinity of the field reversal point $\tilde{x}$ are much less than ε_1. The ionization begins at a certain point $\bar{x} > \tilde{x}$, ends at $x = x_0$ (figure 7.9) and is periodic in $e\varphi$ with a period equal to ε_1 [5]. Such spatially periodic plasma sources are responsible for a periodic field profile in the positive column that maintains the EDF periodicity in ε owing to the mechanism of the EDF bunching [30]. As a result, the periodic field profile and the EDF periodic in ε maintain each other, producing a stable periodic solution. A similar effect may arise at $L_F \ll \lambda_\varepsilon$, because the intermediate electron sources may greatly differ in their energies.

Another scenario of the reconstruction of the dc field and the formation of a positive column will occur at elevated pressures, when the particle loss in the Faraday space is due to the bulk recombination and the discharge gap is longer than that given by equation (7.48). The second field reversal and the transition to the column should be expected at $x > \Lambda_f + l_r$. This length is small; so the cathode region is likely to be short, $\sim(10^{-1} - 10^{-2})$ cm. However, discharge experiments in pumped nitrogen [31] show that an abnormally long, from about 1 to 3 cm, Faraday space may arise here. This paradox can be interpreted as follows. When the collision frequency is energy-dependent, charged particles are transported via the ambipolar drift [32] because of the temperature (field) dependence of the electron mobility. The drift velocity of electrons is about that of ions and is directed to the anode if the electron mobility decreases with the field rise, as is the case in nitrogen; otherwise it is directed to the cathode. This is why the plasma density profile in the Faraday space at $x > \bar{x}$ has a large scale, $l_E \sim b_i E \tau_r \gg l_r$. There are no trapped electrons in most of the Faraday space, so the fluid model holds true. From equation (7.33), we have

$$D_a n'' - (b_i E n + U n)' + \nu_i(E)n - n/\tau = 0,$$
$$e E b_e(E) n = j. \tag{7.61}$$

The right-hand side of the upper equation (7.61) has an additional term to account for the neutral gas pumping velocity U relative to the cathode. By substituting the function $E(n)$ from the other equation and ignoring the diffusion, we obtain

$$(V(E(n)n) + U)n' = n(\nu_i(E(n)) - 1/\tau), \tag{7.62}$$

where the ambipolar drift velocity is $V = \frac{d}{dn}(b_i E n)$. Equation (7.62) describes a monotonic fall of the plasma density to the value n_c for the positive column:

$$\nu_i(E(n)) - 1/\tau = 0.$$

It is worth noting that the Faraday space shape at high and moderate pressures follows closely the $\nu(v)$ dependence, so one should expect a short Faraday space length in gases with a rapidly falling curve $\nu(v)$.

A glow discharge in a gas flow [33] is used to produce a uniform non-equilibrium medium at elevated pressures. Because most of the Joule heat is transferred from the electrons to the neutral gas and the heat removal to the walls is hampered at higher pressures, the difference between the gas temperature and the electron energy decreases, making the plasma closer to equilibrium. The pumping carries out the hot neutral gas, increasing the plasma non-equilibrium. Although a gas flow discharge has some specificity in suppressing instabilities and related non-uniformities, it does not, in principle, differ from the discharges discussed above. There is, however, one thing that is worth noting. If the ambipolar drift velocity is directed from the cathode to the anode, the flow in the same direction increases the path, along which the perturbations are attenuated according to equation (7.62). A reverse flow may lead to a nearly complete disappearance of the Faraday dark space [32, 33].

In discharges with heated (thermal emission) cathodes, the potential fall is of the order of the gas ionization potential. This major feature makes one attribute them to arcs [1]. However, the distributions of luminosity, field and potential in them are quite similar to those shown in figure 5.3. The thin cathode sheath is collisionless, and the outgoing electron flux creates a negative glow region (less intense than with a cold cathode) and a Faraday space with the reverse field.

In closing, we derived similarity relations allowing one to rapidly estimate basic parameters of the short glow discharge and predict them under particular conditions with regard to non-local ionization in the PNG. The effect of such an ionization is shown to be insignificant: more than half the ions from the PNG return to the cathode and contribute to the electron emission. The expressions derived here are not much more complicated than those following from the Engel–Steenbeck classical local model and require knowledge of the same parameters (coefficients α and γ). Relationships to calculate the *IV* characteristic, as well as the distributions of the charged particle concentration along the discharge gap, potential, and electric field (including location of the point of electric field reversal), are presented. It is shown that, for argon, our model gives a much better fit to experimental data than the local models for the near-cathode region of the glow discharge.

7.3 Influence of side walls on near-cathode plasma properties [36]

As noted above, the presence of the side walls plays an important role in the course of various processes and formation of the discharge properties. To a large extent this is due to the non-local nature of the EDF. One of the important effects can be the charging of the side walls to the potentials of much larger than the average electron energy in the bulk plasma as a result of the action of electrons accelerated at the near-cathode jump of potentials (see, reference [34] and references therein). This effect is analogous to charging the walls in the afterglow plasma with a non-local group of energetic electrons produced in some plasma-chemical processes [35]. Below we discuss the influence of the wall on the dimensions of CF, NG and FDS.

In reference [36] to study the influence of the discharge tube radius on the near-cathode plasma properties and discharge structure dimensions, a full-scale 2D modeling of a short discharge in argon has been performed. The simulations were conducted for gas pressure $p = 3$ Torr and the discharge length $L = 1$ cm, but with different tube radii. Besides $R_d = 15$ mm, tube radii of $R_d = 5$ mm and $R_d = 2$ mm were studied. Corresponding *IV*-traces are shown in figure 7.10. For convenience, the *x*-axis shows current densities, averaged over the tube cross-section, $\langle j \rangle_r$, rather than the discharge current since the tube cross-section area changes with radius. The calculated numerical values of the voltages, currents, current densities, cathode

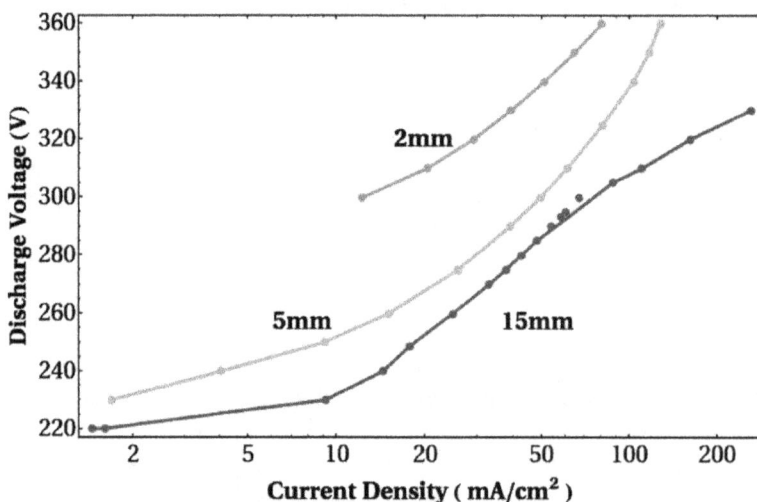

Figure 7.10. *IV*-trace of argon dc discharges for $L = 1$ cm and $R_d = 2$, 5 and 15 mm. From [36].

sheath thicknesses and coefficients of ion utilization (essentially a plasma effectiveness ratio which determines the fraction of the total number of ions generated in the plasma which return to the cathode) for the different tube radii of the IV-traces in figure 7.10 are also given in table 7.2.

Table 7.2. Tube radius, calculated voltage, discharge current, average current density, cathode sheath thicknesses and a coefficient of ion utilization (δ_i). From [36].

R_d, mm	U, V	I, mA	$\langle j \rangle_r$, $mAcm^{-2}$	d_1, mm, equation (7.63)	d_2, mm, equation (7.64)	δ_i
15	220	10.05	1.42	0.88	0.96	0.98
15	230	64.16	9.08	0.75	0.84	0.95
15	240	100.7	14.24	0.68	0.75	0.94
15	249	124.4	17.59	0.63	0.7	0.94
15	260	174.4	24.67	0.59	0.65	0.94
15	270	231.5	32.75	0.55	0.6	0.94
15	275	263.9	37.34	0.53	0.58	0.94
15	280	298.4	42.22	0.51	0.56	0.94
15	185	336	47.53	0.5	0.54	0.94
15	290	376.8	53.31	0.48	0.52	0.94
15	293	406.8	57.55	0.47	0.51	0.94
15	295	421.1	59.57	0.47	0.51	0.94
15	300	468.8	66.31	0.46	0.49	0.94
15	305	642.2	90.86	0.37	0.38	0.96
15	310	859.8	121.6	0.35	0.36	0.96
15	320	1373	194.3	0.32	0.32	0.96
15	330	1649	233.3	0.3	0.31	0.96
5	230	1.28	1.62	0.85	0.89	0.8
5	240	3.06	3.9	0.76	0.86	0.8
5	250	6.88	8.77	0.69	0.8	0.79
5	260	11.36	14.46	0.63	0.73	0.79
5	275	19.47	24.79	0.56	0.65	0.79
5	290	29.26	37.26	0.51	0.58	0.8
5	300	37.05	47.18	0.48	0.55	0.8
5	310	45.62	58.09	0.46	0.52	0.8
5	325	60.33	76.82	0.43	0.48	0.81
5	340	78.12	99.46	0.41	0.45	0.81
5	350	89.59	114.1	0.4	0.43	0.81
5	360	96.93	123.4	0.4	0.41	0.81
2	300	1.51	12	0.62	0.73	0.47
2	310	2.5	19.93	0.55	0.66	0.5
2	320	3.59	28.53	0.51	0.62	0.52
2	330	4.81	38.27	0.49	0.57	0.53
2	340	6.25	49.72	0.46	0.54	0.54
2	350	7.93	63.09	0.44	0.51	0.55
2	360	9.85	78.42	0.42	0.48	0.56

It can be seen that the *IV*-traces are displaced to higher voltages with decreasing tube radius. The discharge voltage increases with decreasing tube radius, for a constant current density, $\langle j \rangle_r$, while the current density increases with tube radius at a constant discharge voltage. The decrease in tube radius leads to an increase in wall losses of charged particles requiring an increase in the discharge voltage to maintain the same current density.

The 2D ion density distributions at a constant voltage of $U = 300$ V for $R_d = 15$ mm (a), $R_d = 5$ mm (b) and $R_d = 2$ mm (c) are shown in figure 7.11. Due to the symmetry, only half of the tube radial dimension is shown in the figure. Axial electron and ion density distributions, n_e (red) and n_i (blue), respectively, on the axis of the discharge with $\langle j \rangle_r = 48$ mA cm^{-2} and different tube radii are shown in figures 7.12(a)–(c).

The axial profiles in figure 7.12 are typical for abnormal discharges. The cathode of the abnormal discharge is completely filled with the cathode sheath near the cathode at distance d. The distance d (denoted as d_1) was found from the condition where

$$n_e(d_1) = 0.5 n_i(d_1). \tag{7.63}$$

Almost the entire applied potential drop U is in the cathode sheath. With the cathode sheath so close to the cathode, the ions transverse displacement (Δr) for their drift time to the cathode, in the strong electric field of the cathode sheath,

$$\frac{\Delta r}{d} \approx \sqrt{\frac{\lambda_i}{d}} \ll 1,$$

is negligible. Here, λ_i is the mean free path of an ion. Thus, all the ions move in the cathode sheath toward the cathode along horizontal flow lines, even in tubes of small radius and all ions, from the cathode sheath to the cathode, return to the cathode with negligible wall loss. Therefore, a 1D solution of the Poisson equation is sufficient to estimate the cathode sheath parameters. Since the cathode sheath electric field is strong, the square root dependence of the ion drift velocity ($v_i = k_i\sqrt{E}$) is more appropriate, and the current density can be expressed as

$$\frac{j}{p^2} = \frac{\sqrt{2}}{3\pi} \frac{(k_i p) U^{1.5}}{(pd)^{2.5}} A \text{ (cm Torr)}^{-2}.$$

For an Ar discharge at the pressure $p = 3$ Torr, the average current density is

$$\langle j \rangle \approx 2330 U^{1.5}/d_2^{2.5} \text{mA cm}^{-2}, \tag{7.64}$$

where d_2 is the cathode sheath thickness in centimeters. The cathode sheath thickness d_2 from equation (7.64) is given in table 7.2 and is seen to be in good agreement with the numerical simulations. The above analysis shows that even in a 2D model, the processes in the cathode sheath are determined principally by their longitudinal coordinate and thus a 1D model is sufficient to determine the cathode sheath parameters.

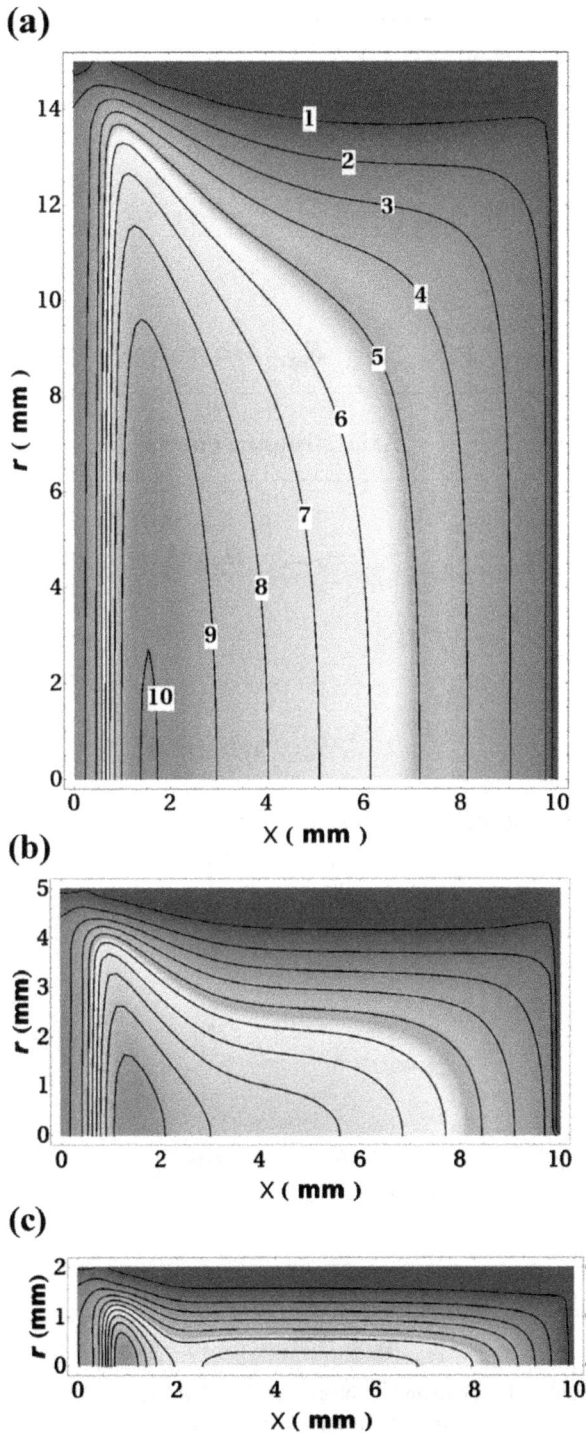

Figure 7.11. 2D profiles of positive ion density of at $U = 300$ V for $R_d = 15$ mm (a), 5 mm (b) and 2 mm (c). The isolines from 1 to 10 on plots (a), (b) and (c) correspond to increasing the plasma density from 2.5×10^{10} (1) to 2.5×10^{11} (10) cm^{-3} with increments of 2.5×10^{10} cm^{-3}. From [36].

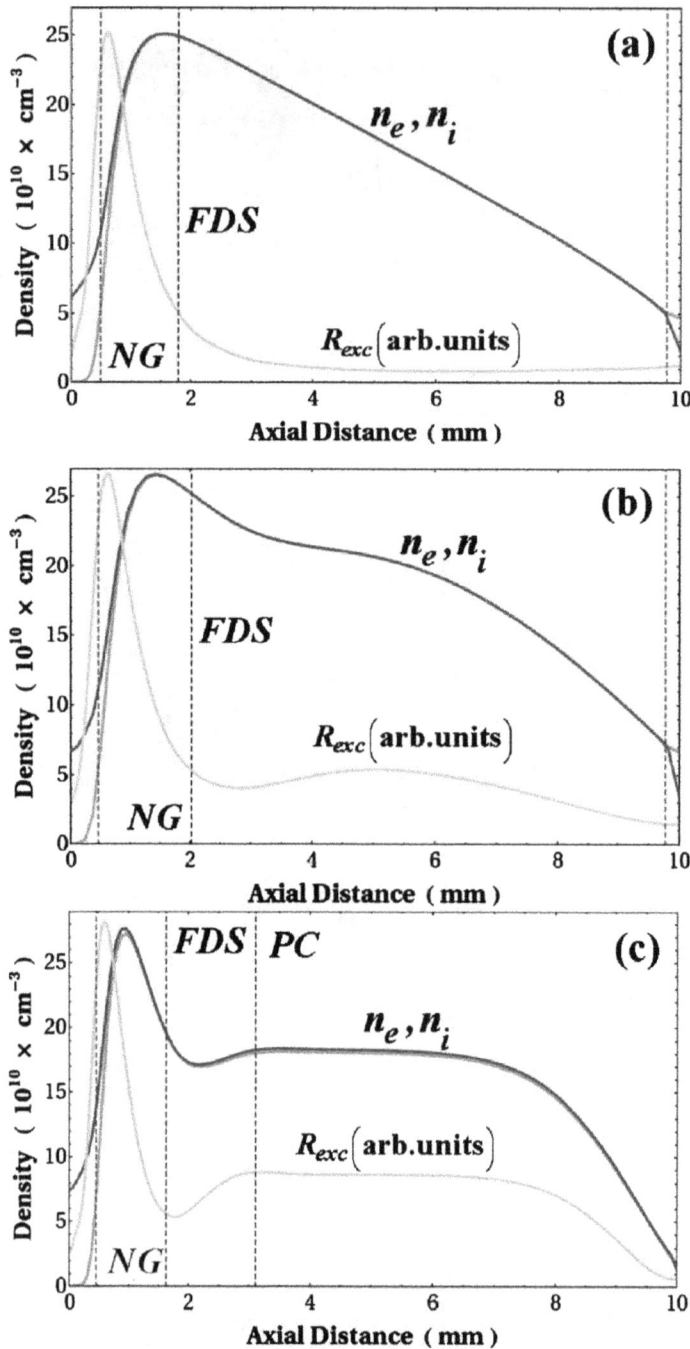

Figure 7.12. Axial of profiles of n_e (red) and n_i (blue) densities of with discharge current $j = 48$ mA cm^{-2} for $R_d =$ (a) 15 mm, (b) 5 mm and (c) 2 mm. Also, the axial profiles of excitation rate R_{exc} are shown in arbitrary units. Vertical dashed lines show the borders of discharge regions from left to right: CF, NG, FDS, and AF for (a) and (b) or PC for (c). From [36].

The different situation with other regions of the discharge. From figures 7.11 and 7.12, it can be seen that the density of charged particles, on the central axis, peaks near the cathode sheath then decreases smoothly toward the anode. The ion density profile in the quasi-neutral discharge region of a single-component plasma can be found from an ambipolar diffusion equation.

Analysis of reference [36] shows that for $R_d > 2.4L/\pi$ (or $R_d/L > 0.76$) the radial loss on the plasma density balance is small. The plasma density is almost independent of the transverse dimension which results in a radial distribution that is nearly constant in the central region and decreases only near the wall. That is seen for the $R_d = 15$ mm radial profile in figure 7.13, which shows the charged particle radial density distribution with $\langle j \rangle_r = 48$ mA cm^{-2} for $R_d = 15$ mm (a), $R_d = 5$ mm (b) and $R_d = 2$ mm (c) at the axial position of maximum plasma density for each tube radius.

Therefore, for cases when $R_d > 2.4L/\pi$, a 1D model should give an adequate description of the dc discharge. However, as the tube radius decreases, radial losses begin to play a notable role. The radial and longitudinal spatial distributions of the charged particle densities for a tube of radius of 5 mm (figures 7.12(b) and 7.13(b)) change if the spatial distribution is larger than for 15 mm tube (figures 7.13(a) and 7.12(a)). The higher charged particle losses, that accompany the tube radius decrease, decreases the longitudinal dimensions of the negative glow and Faraday dark space regions. To sustain the discharge, the positive column arises as the negative glow and Faraday dark space dimensions decrease in figure 7.12(c).

The radial distributions similarly change with decreasing radius. Figures 7.13(b) and 7.13(c) show the radial distribution for $R_d = 5$ and $R_d = 2$ mm, respectively. The distributions decrease more gradually with radial position than the 15 mm tube, i.e. the density roll-off is more gradual resulting in a more peaked distribution becoming approximately Bessel-like at $R_d = 2$ mm. Similar changes in the charged particle density distributions are observed for fixed voltage in figure 7.11. Thus, for the tube with radius comparable to the length wall effects start to impact charged particle density distributions.

For discharges with radius $R_d < 0.76L$ wall effects cause deformation of the longitudinal and radial spatial density distributions and charged particle flows in the plasma. This can lead to a number of unexpected phenomena in such a self-consistent system as a gas discharge. Since the charged particle density distribution depends on radial position, the cathode sheath thickness, obtained from the condition $n_e(d) = 0.5n_i(d)$, will depend on radial position. This sufficiently paradoxical fact is illustrated in figure 7.14, which shows the cathode sheath thickness for $R_d = 15$ mm (a), $R_d = 5$ mm (b) and $R_d = 2$ mm (c) tube radius, at constant average current density, $j = 48$ mA cm^{-2}.

Thus, the cathode sheath thickness, d, in the glow discharge does not have a constant value over the cross-section, but depends on the radius. Furthermore, though the lines of ion current in the cathode sheath are horizontal, it does not mean that the radial density distribution and ion current in the cathode sheath and on the cathode surface are transversely uniform (as is assumed in usual 1D models). The absolute value of current lines in the cathode sheath, as in the plasma, decreases

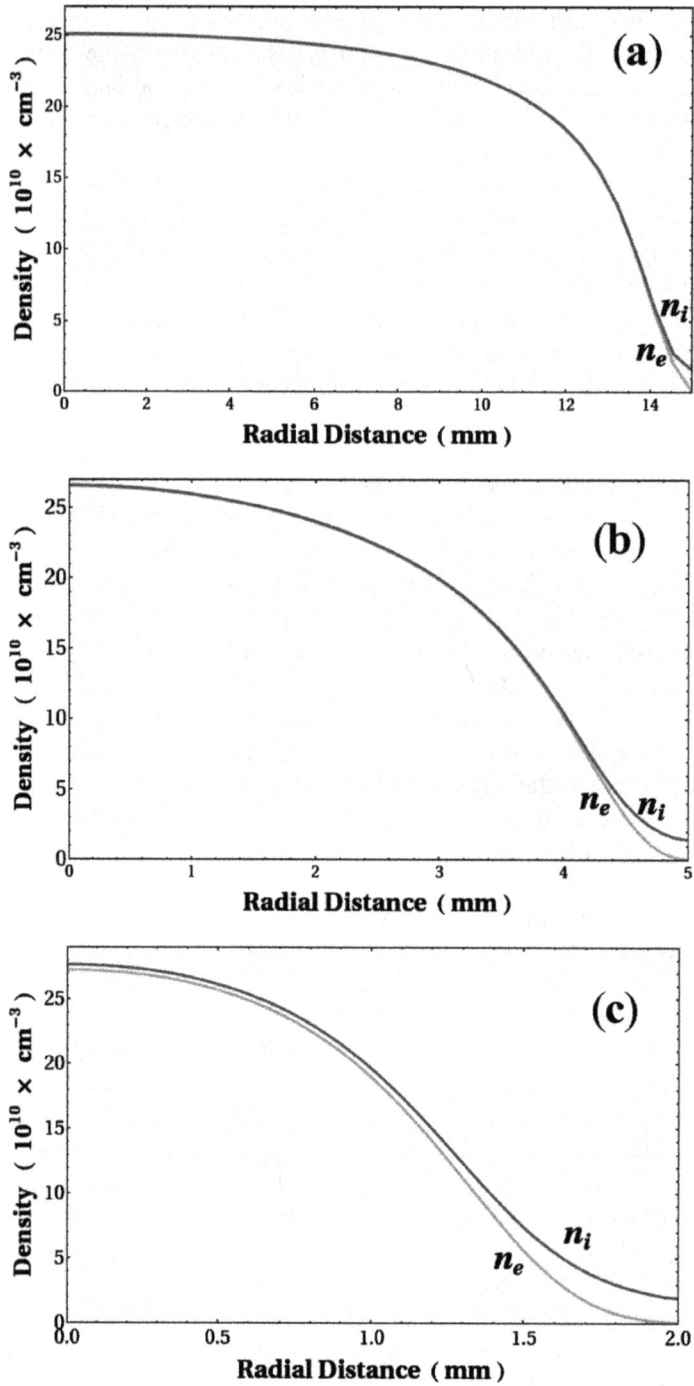

Figure 7.13. Radial profiles of electron and positive ion densities at axial coordinates of maximum charged particle density with $<j> = 48$ mA cm^{-2} for $R_d = 15$ mm (a), 5 mm (b) and 2 mm (c). From [36].

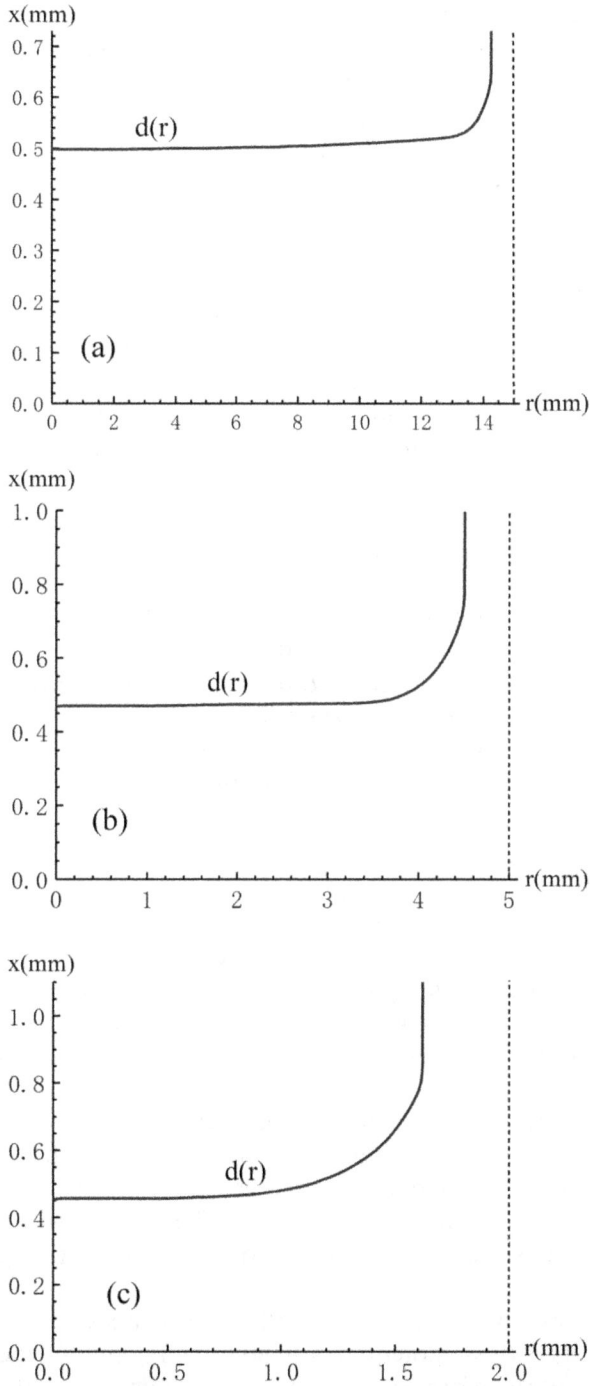

Figure 7.14. Cathode sheath thickness radial profile for $R_d = 15$ mm (a), 5 mm (b) and 2 mm (c) with $<j> = 48$ mA cm^{-2}. The dashed vertical line shows the tube position. From [36].

from the center axis of the discharge to the periphery. In turn, this must lead to transverse inhomogeneity of the ion current density on the cathode surface which is shown in figure 7.15 for constant voltage, $U = 300$ V.

Thus, the results of simulations [36] and the above analysis show that the 2D profiles of the parameters of the region near the cathode of an abnormal dc glow discharge depend on the relationship between the cathode sheath thickness, d, the length of the discharge volume, L, the tube radius, R_d, and the characteristic length of ionization decrease in the negative glow plasma, λ_i. Three cases have been found (assuming that coefficient 0.76 above may be replaced by unity).

(1) For $R_d > L$, the results of simulations with a 1D model give satisfactory accuracy. An analytical 1D model [37], which takes into account non-local ionization in the plasma of the negative glow is adequate to determine the fundamental characteristics of an abnormal dc glow discharge. However, the cathode sheath cannot be assumed to be self-contained and independent of the plasma and a condition for the self-maintenance of the discharge on the basis of local models is not applicable.

(2) For the tube radius comparable to or just smaller than the discharge gap, $R_d \leqslant L$, the loss of charged particles to the walls significantly affects not only the fundamental characteristics of the plasma near the cathode, but also the cathode sheath. If $\lambda_i < R_d$, practically all the ions produced in the negative glow plasma due to non-local ionization, return to the cathode and coefficient of ion utilization $\delta_i < 1$.

(3) For long narrow discharges $R_d < L$ on the right side of the Paschen curve, the ion utilization coefficient, $\delta_i < 1$, will be small when $R_d < \lambda_i$. It should be noted that in this case, the cathode sheath can be considered nearly self-contained and the local condition of self-maintenance in the form of von Engel–Steenbeck is acceptable.

7.4 Active control of plasma properties with application of external voltage to the side wall [38, 39]

This section demonstrates the application of external potential to the side wall (active boundary) for controlling plasma properties. To demonstrate those effects in plasma of short dc discharge with cold cathode, modeling the discharge with application of different voltages to the discharge wall has been performed in [38]. The modeling has been performed in argon gas discharge with 1 Torr pressure. The length of the gas chamber was taken as 12 mm and radius was 12.5 mm. The wall was divided into three parts. The central part was metallic from 0.5 to 11.5 mm with a constant potential V_w. The side fragments (near the cathode and anode) were taken as dielectric ones with dimensions of 0.5 mm. Their potential can be different in different points of the surfaces. This model corresponds to the experimental device shown in figure 7.16.

Three regimes have been modeled. In all three regimes, the anode potential is equal to 0 V and the cathode potential is −180 V. In regime 'A' the wall potential is also equal to 0 V and therefore coincides with the anode potential. In regime 'B' the wall potential is −25 V. In the last regime the 'C' wall potential is −50 V. As an example of calculations, a typical result of 2D modeling of argon metastable density

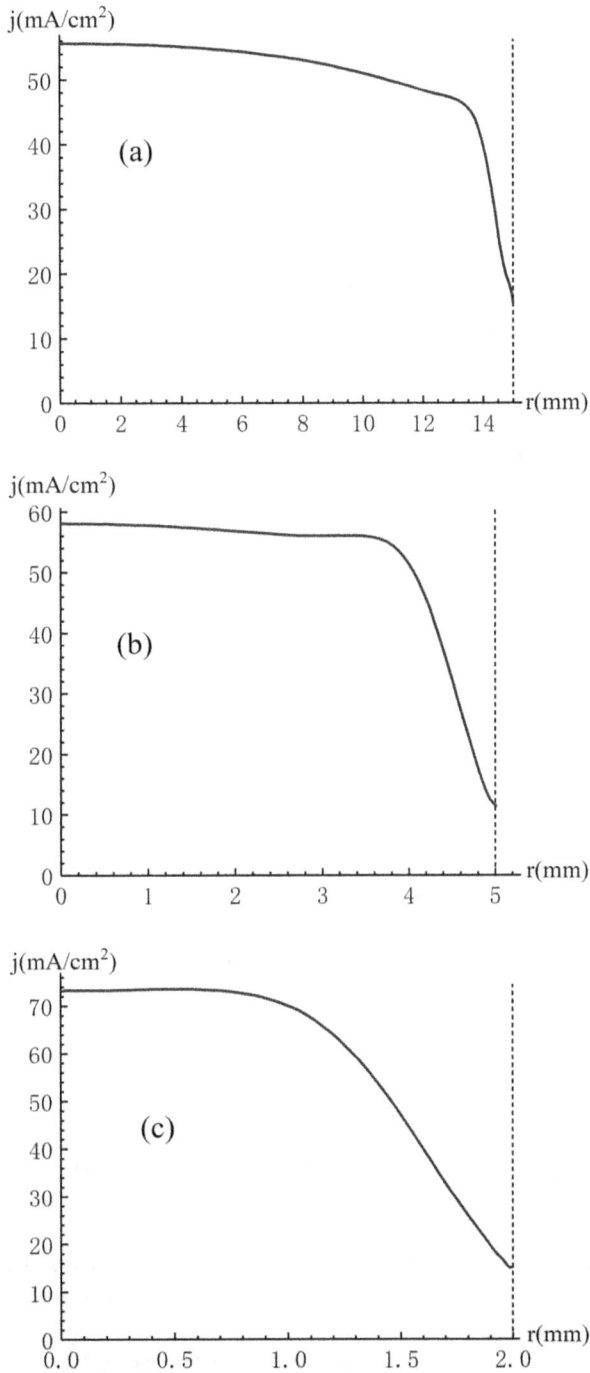

Figure 7.15. Current density radial profile on the cathode for $U = 300$ V. $R_d = 15$ mm (a), 5 mm (b) and 2 mm (c). The dashed vertical line shows the tube position. From [36].

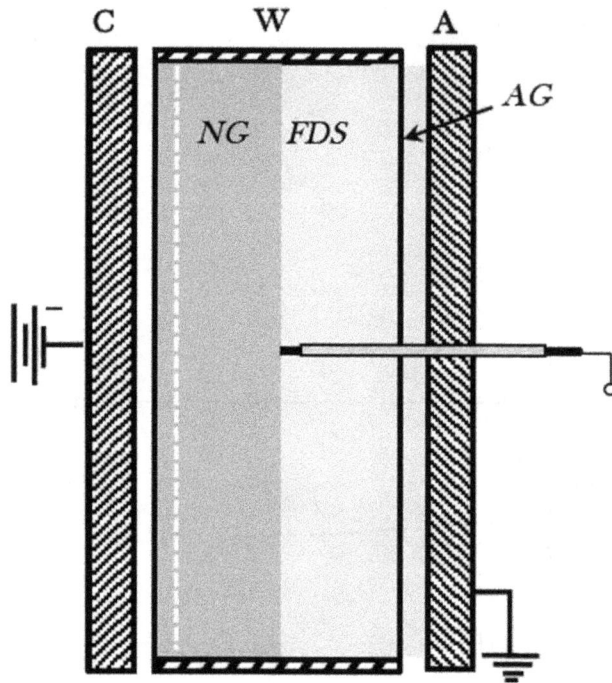

Figure 7.16. Schematic diagram of an experimental device consisting of cathode C, anode A, and cylindrical wall W. Typical structures of the discharge plasma include the negative glow NG and anode glow AG, respectively, and Faraday dark space FDS. The cathode sheath boundary, located in the NG, is indicated by the dashed line. From [38].

is shown in figures 7.17. In those figures, the top plot corresponds to regime 'A', the middle plot corresponds to regime 'B', and the bottom plot corresponds to regime 'C'. To make the modification of plasma properties more visible, figures 7.18, 7.19 and 7.20 show the comparison of axial plasma properties for the different regimes.

It is possible to see from figures 7.2 and 7.18 that applying more negative voltage to the wall does not significantly change argon metastable-atom density. In contrast, figure 7.4 shows that slow, thermal electron density (the same for ions) is significantly reduced by applying increasingly more negative potential to the walls. Figure 7.5 shows considerable increasing electron temperature in the plasma volume with more negative wall voltage. Interpreted together, the ratio between slow, thermal electron (ion) density and metastable-atom density decreases with increasing negative wall voltage and increasing slow-electron temperature.

An explanation of the observed phenomenon is as follows. Energetic electrons leave the cathode sheath with the energy of about 180 eV and diffuse in the direction of the anode and walls while simultaneously ionizing and exciting metastable states of neutral atoms in the volume. During this process, they reduce their energy and create a continuous electron spectrum of energetic electrons at EDF. Only the energetic electrons with energies $\varepsilon > eV_w$ can reach the walls. Less energetic electrons can go to the anode only. Therefore, as the wall potential is essentially

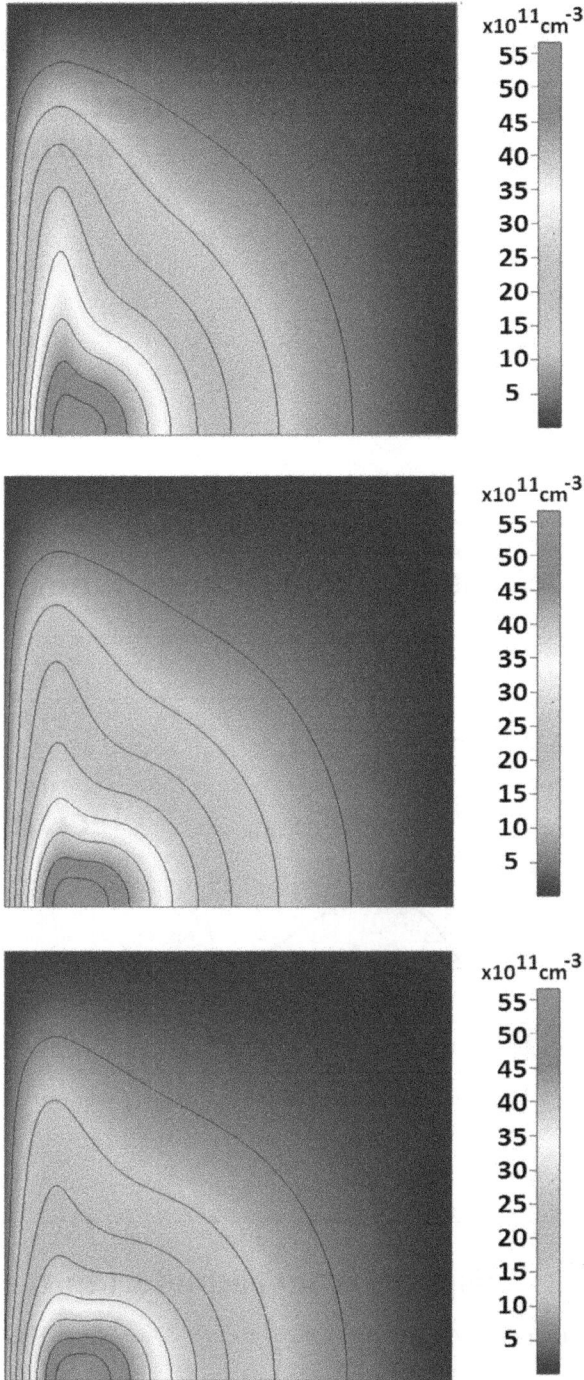

Figure 7.17. 2D distribution of the metastable-atom density (in the units of 10^{11} cm^{-3}). Wall potential is 0 V (top, regime A), 25 V (middle, regime B) and 50 V (bottom, regime C). From [38].

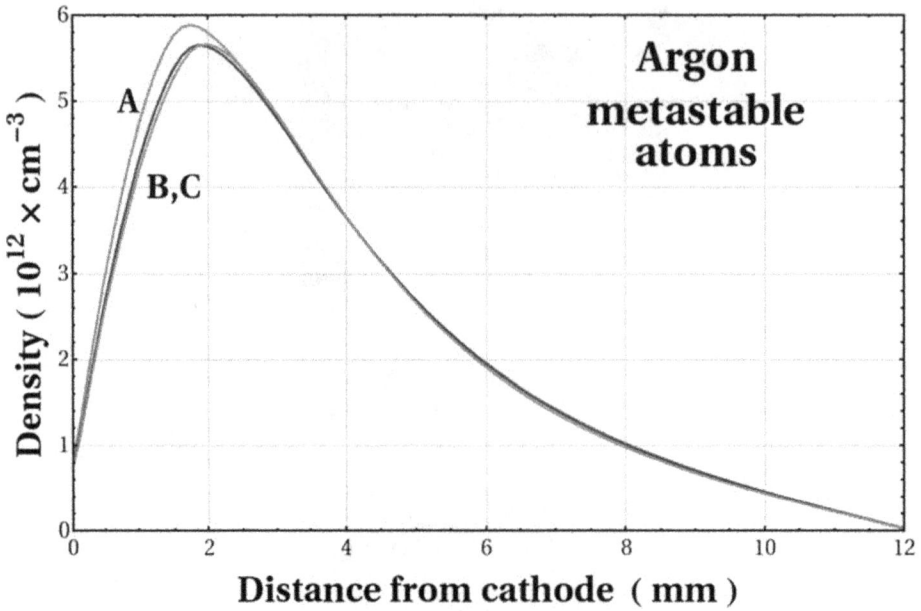

Figure 7.18. Axial distributions of metastable density. [38].

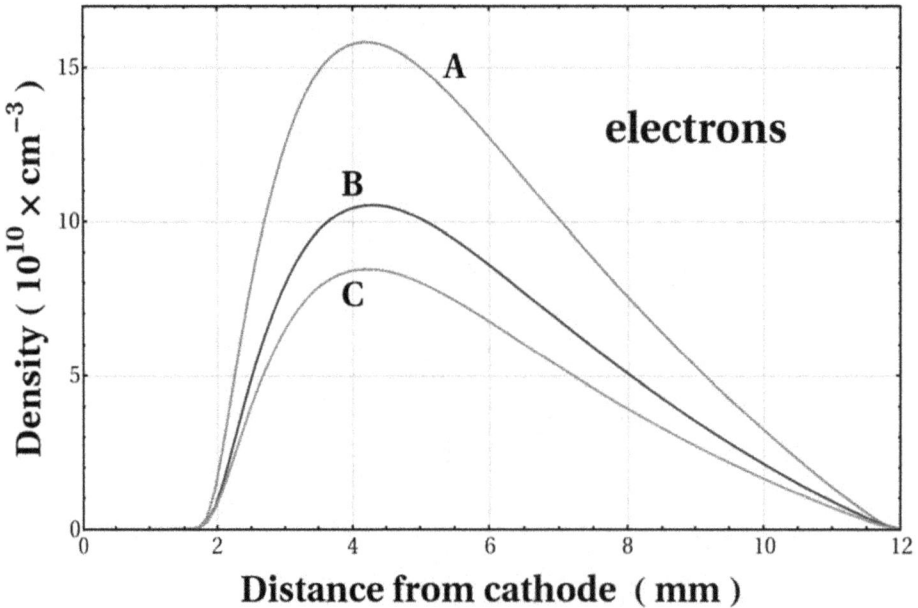

Figure 7.19. Axial distributions of electron density. From [38].

less than the energy of the most energetic electrons, the negative voltage V_w, applied to the conducting (active) walls, can modify significantly the trapping of the low-energy part of energetic electrons (the minority of the energetic electrons), while not

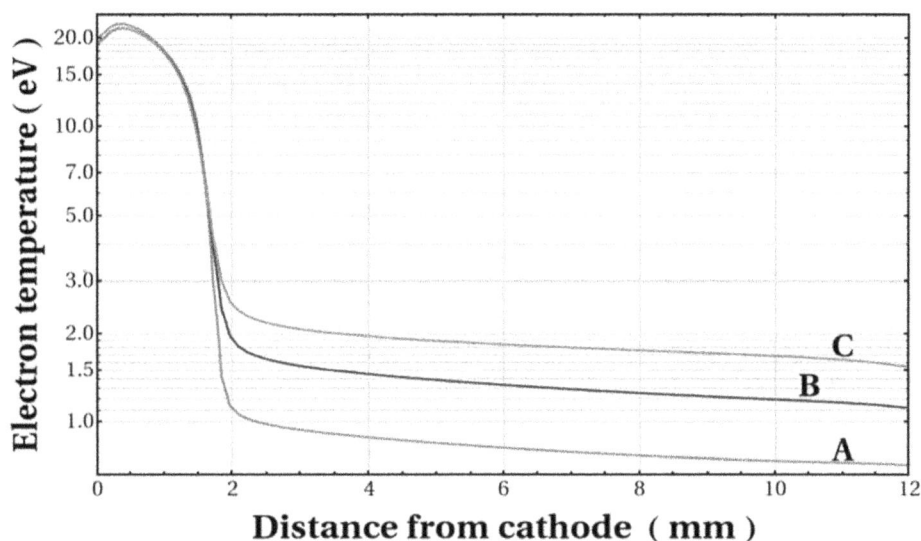

Figure 7.20. Axial distributions of electron temperature. From [38].

essentially changing the high-energy (majority) part of those electrons. As a result, the excitation and ionization production depends only weakly on the applied potential to the wall. Metastable atoms disappear at the walls and during binary collisions (Penning ionization), independent of the wall potential. Therefore, the density of metastable atoms should not depend on negative wall potential, as confirmed in figures 7.2 and 7.18.

Similarly, the production of ions (slow, thermal electrons) depends weakly on the negative voltage applied to the walls. However, the heating of slow electrons is mostly due to the low-energy part of energetic electrons, as their collision frequency with slow electrons is nearly inversely proportional to $\varepsilon^{1.5}$. At the same time, varying the negative wall potential modifies particularly the low-energy part of the energetic electrons and changes the heating of the slow, thermal plasma electrons by the energetic electrons (as the heating is mostly due to the low-energy part of those electrons), which consequently changes the electron temperature and their diffusion rate to the anode. Therefore, due to the additional heating of the slow electrons (see figure 7.5), their diffusion to the anode will be faster and their density goes down with increasing negative voltage applied to the walls. As a result, while the density of metastable atoms depends weakly on wall voltage, the density of slow electrons (ions) depends strongly on wall voltage, significantly reducing with increasing negative wall voltage. Figure 7.21 shows both behaviors more clearly.

Reference [39] provides some confirmation to the above modeling. A typical result of the experiments measuring the electron temperature T_e as a function of applied wall potential with respect to the wall voltage (using argon gas, a discharge current of 5 mA and gas pressure of 2.2 Torr) is provided in figure 7.22. The application of additional negative wall voltage leads to an increase in the electron

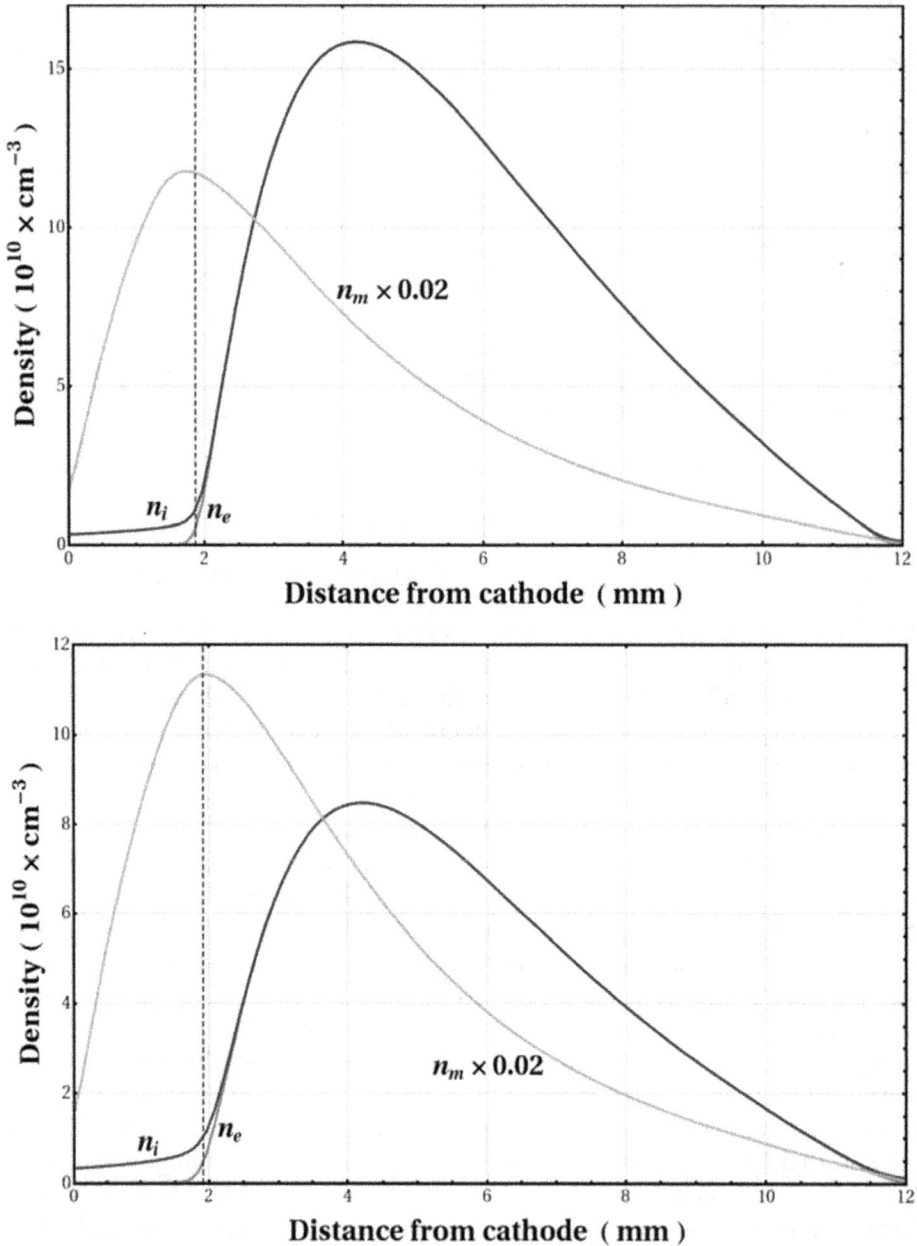

Figure 7.21. Axial distributions of densities of slow electrons, ions, and metastable atoms for regimes A (top) and C (bottom). From [38].

temperatures caused by the partial trapping of the energetic electrons (increasing the lifetime of the energetic electrons in the plasma volume) and energy transferring collisions being heated by slower electrons. Consequently, it is possible to increase the electron temperature by increasing the absolute value of the applied negative

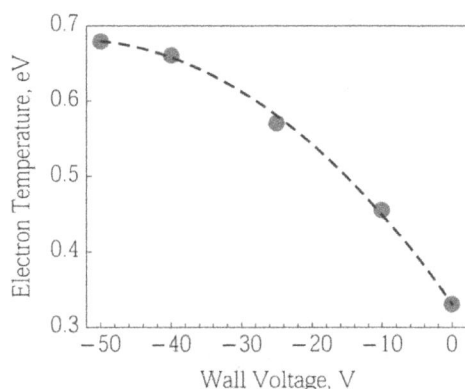

Figure 7.22. Dependence of electron temperature on wall voltage in argon discharge with current of 5 mA and gas pressure of 2.2 Torr. From [39].

potential. Because this process depends primarily on the applied voltage, it is possible to have a regulation of the electron temperature.

Thus, the application of negative voltage to the discharge walls could change the trapping of the low-energy part of the energetic electrons that are emitted from the cathode sheath and that arise from the atomic and molecular processes in the plasma within the device volume. The low-energy part of the energetic electrons is responsible for heating the slow, thermal electrons. At the same time, the production of slow electrons and metastable atoms is mostly due to energetic electrons with higher energies. The variation of electron temperature results in a changing decay rate of slow, thermal electrons, while the decay rate of metastable atoms and production rates of slow electrons and metastable atoms are practically unchanged. The ability to control the electron/metastable density ratio and the electron temperature represents an important capability and is the main result here.

7.5 Is negative glow plasma of a direct current glow discharge negatively charged? [40]

It is well-known, that in any bounded plasmas, when electrons and positive ions are lost at the boundaries, the diffusive fluxes of the charged particles are directed from the plasma center toward the walls. In this situation, the ambipolar electric field has to decelerate the electrons, aligning their flux with the positive ion flux. To create such an electric field, the charge density of the plasma must be positive. At the same time, almost all textbooks on gas discharge physics indicate that the negative glow and/or Faraday dark space plasmas of a dc glow discharge are negatively charged. Furthermore, the charge separation there is even stronger than in the positive column (see, e.g. figure 14.1 in [41]). This statement is in contradiction with the above consideration that in simple plasma charge must be positive. Furthermore, because the density of charged particles in the NG plasma exceeds the corresponding density in the PC region, it is difficult to expect greater deviation from the quasi-neutrality anywhere in the NG than in a PC region.

This issue was studied in detail in [40]. To do that the distributions of the potential (φ) and the electric fields (E_x) along the discharge gap (x) should be considered as the magnitude and sign of the space charge (ρ) is determined by the Poisson equation. Then, the 1D model of the discharge predicts a negative space charge in the NG [41]. However, in this situation, it is unclear what physical mechanisms may provide the plasma quasi-neutrality. To resolve this situation it was shown in [40], that an adequate description of diffusive glow discharges is possible only in 2D/3D models (the 1D model gives the solution only in the presence of volume recombination).

To demonstrate the above considerations, the 2D numerical simulation of the dc glow discharge in helium in a cylindrical tube has been performed. The results of the simulation that demonstrates the positive charge density in the NG region are shown in figure 7.23. At the gas pressure of $p = 1$ Torr and $U_0 = 1$ kV, an abnormal glow discharge with the current $I = 982$ μA was obtained. As observed from figure 7.23, the gap is long enough and contained all of the main parts of a dc glow discharge: CF, NG, FDS, PC and AF. The PC in this case is quite short, but pronounced. The vertical dashed lines at figure 7.23 (as well as figures 7.24 and 7.25) indicate the boundaries of the main discharge areas. The NG and FDS regions are joined in the figures. The radial profiles of the electron density are similar to the Bessel profiles both in the PC and NG regions. The $n_i n_e$ profile, which is proportional to ρ, is presented in figure 7.23. It can be observed that $\rho(x, 0)$ is positive for all x, as expected.

Figure 7.24 shows axial profiles of electric potential and axial component of electric field. It can be observed that $\varphi(x, 0)$ had an inflection point, and $E_x(x, 0)$ had one decreasing interval and two increasing intervals. These profiles were very

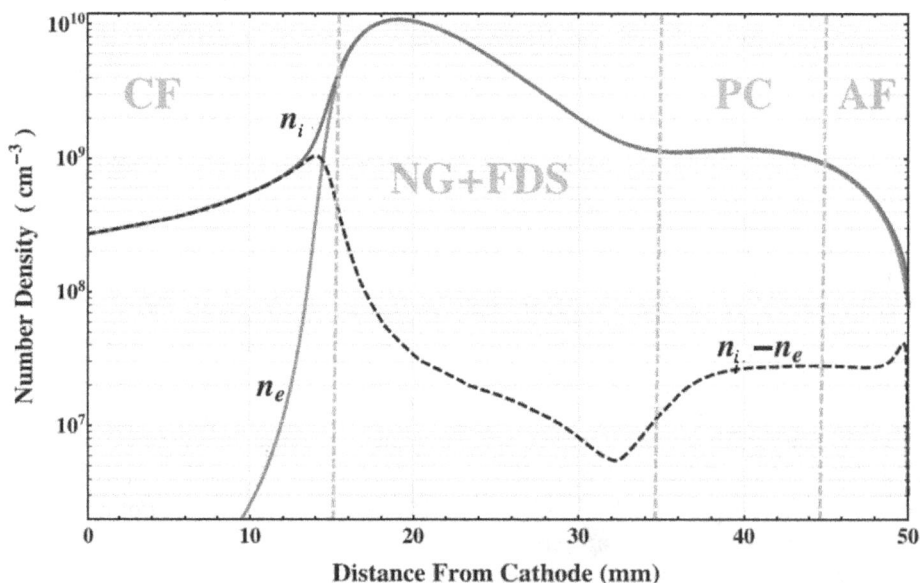

Figure 7.23. Axial profiles of the electron density n_e, ion density n_i, and their difference $n_i n_e$. In this and subsequent figures, the vertical dashed lines mark the main discharge regions. From [40].

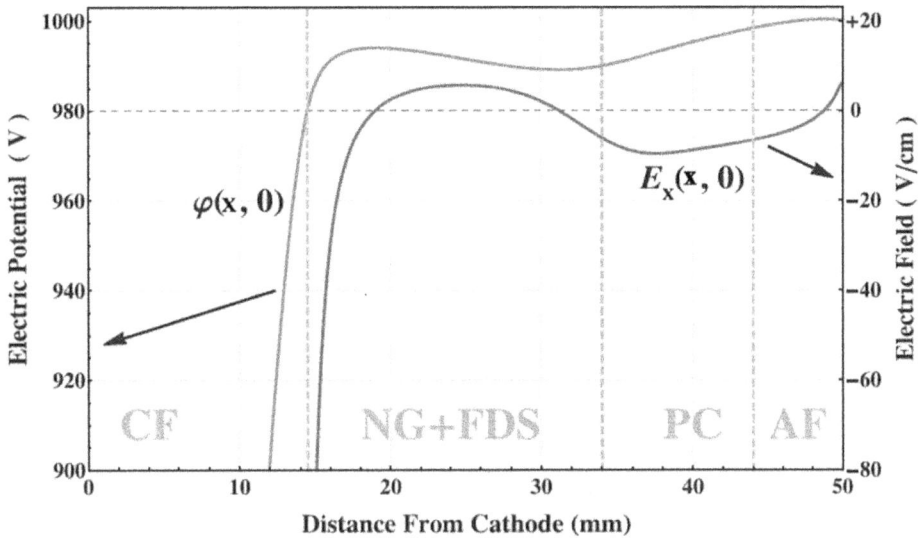

Figure 7.24. Axial profiles of electric potential (left scale) and axial component of electric field $E_x(x, 0)$ (right scale). From [40].

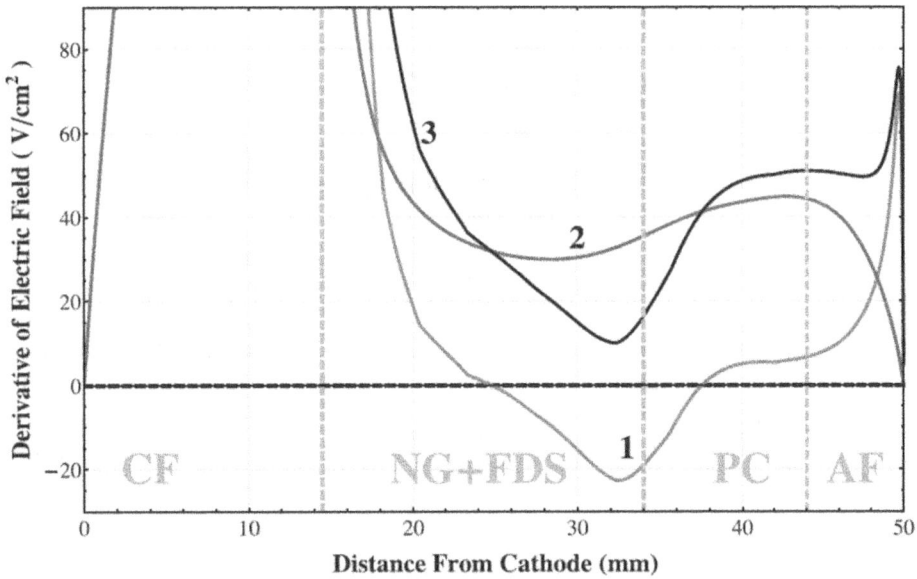

Figure 7.25. Axial profiles of the terms in the Poisson equation (2): $\partial E_x/\partial x$ (1), $(1/r)\partial(rE_r)/\partial r$ (2), and charge density ρ/ε_0 (3). From [40].

similar to those that are often cited in the literature on the classical dc glow discharge in a tube. Figure 7.25 shows the axial profiles of the terms $\partial E_x/\partial x$, $(1/r)\partial(rE_r)/r$ and ρ/ε_0 in the Poisson equation

$$\partial E_x/\partial x + (1/r)\partial(rE_r)/r = \rho/\varepsilon_0/\varepsilon_0$$

in the plasma (NG, FDS, and PC regions). It can be observed that $(1/r)\partial(rE_r)/r > 0$ at all x, $\partial E_x/\partial$ changes sign, and $|(1/r)\partial(rE_r)/r| > |\partial E_x/\partial x|$ at points, where $\partial E_x/\partial x < 0$. Thus, $\rho(x, 0) > 0$ at all x, whereas the 1D model of discharge predicts that $\rho < 0$, where $\partial E_x/\partial x < 0$. The calculations showed that $\rho > 0$ not only on the discharge axis but also in the whole plasma region. This result is due to a monotonic increase of $(1/r)\partial(rE_r)/r$ with respect to r at fixed x.

Thus, in the investigated conditions, the discharge plasma had a positive charge, including the NG plasma, whereas the axial dependence of the axial electric field is non-monotonic. The traditional interpretation that states a negative charge of the NG plasma is based on analogies with a simple 1D model of glow discharge, whereas the actual discharges are always 2D. The radial term in divergence with the electric field can provide a positive charge density. It is demonstrated that the fact that space charge is negative in such classical objects as the NG plasma has no convincing evidence.

7.6 The anode region

In contrast to the cathode, there is no ion emission at the anode, $j_i = 0$. The ion current in the anode region varies from zero to the value in the positive column, which is less than one percent of the total current. The net ionization rate in the anode region is two or three orders of magnitude lower than in the negative glow. As a result, the voltage in the anode region is low, less than 10 eV and the luminosity in it is essentially less pronounced than in the PC. Nevertheless, phenomena occurring in the anode vicinity are as important for the discharge maintenance and stability as those in the cathode region. The experimental work on them, however, has not been as intensive, so our current understanding of the physical processes in it is far from satisfactory. Even the physical reasons, which determine the sign of the anode potential fall, still remain unclear.

There are two main scenarios for the events on the anode. One dates back to Langmuir [42]. It states the following: since the electron current, j_e, in the plasma remains practically constant in the anode region, it is much lower than the thermal (random) current. It would be quite natural to expect that the electric field in the anode sheath hinders the electron motion, and the negative anode potential fall, $\varphi_0 < 0$, is a few times larger than T_e/e. The random current of thermal Maxwellian electrons is described as

$$j_e^{(th)} = en_e\left(\frac{8T_e}{\pi m}\right)^{1/2},$$

and the value of $\varphi_0 < 0$ is determined by

$$j = j_e^{(th)}\exp(e\varphi_0/T_e).$$

The Langmuir model implies the presence of the ion space charge in the anode sheath. It leads to the field reversal and the formation of a potential well for electrons. The ion flux at the anode is directed to the anode. A population of cold trapped electrons similar to that in the negative glow arises in the anode vicinity. To the best of our knowledge, there has been no consistent calculation of the plasma parameters in this model.

The other model is based on the fact that the escape of fast electrons to the anode would distort the Maxwellian EDF, making the problem essentially kinetic even at relatively high plasma densities. Von Engel believed that the solution to this problem could be found with a monotonic potential profile [43]. In that case, the ion flux will rise monotonically with distance to the zero value at the anode surface. The plasma is, therefore, created due to the ions generated in a thin space charge sheath of the anode. The ionization here must be far more intense than the average ionization in the column. The anode fall φ_0 is positive and its value is determined by a sufficient number of ions produced in the sheath in order to generate the ion flux Γ_i necessary for bringing a quasi-neutral plasma to life. To produce the necessary ion flux in the close vicinity of the anode, the potential fall must be of the order of the ionization potential ε_i. The space charge will then be created by electrons, in contrast to the cathode fall.

This problem was investigated experimentally by Klarfeld [44]. Negative φ_0 values and a smooth profile of the plasma glow were observed at high currents (see also [42]), whereas at low currents φ_0 was positive. A thin bright layer could be seen immediately at the anode surface, which was interpreted as being due to a more intense ionization in the sheath. These results are illustrated in figure 7.26. A lower pressure and a smaller anode area were also found to stimulate the transition from positive to negative φ_0 [44]. The physical mechanism of these phenomena still remains obscure. Unclear is also the mechanism responsible for a normal current density across the anode [45]. What is quite clear, however, is that the structure and even the sign of the anode fall are determined by the subtle characteristics of the ionization kinetics.

An attempt to construct a kinetic sheath model of the positive anode potential fall in the collisionless case was made in [46]. The anode sheath in this case is thin with

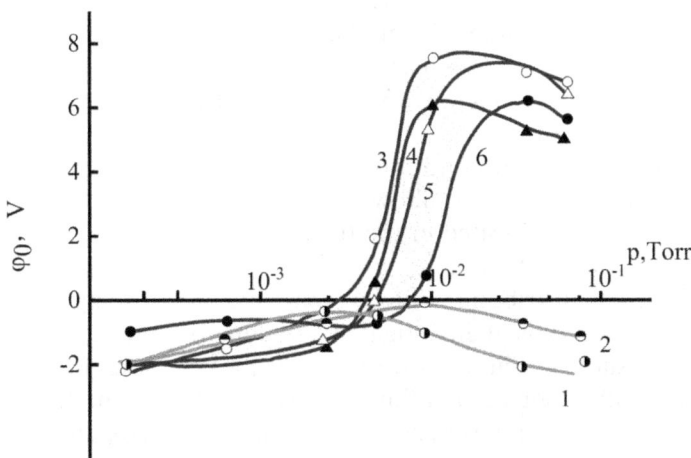

Figure 7.26. Anode fall in Hg discharge. Tube radius is 1.5 cm. Current is 10 A (1), 3 A (2), 0.3 A (3), 1 A (4), 0.1 A (5) and 0.05 A (6). From [15].

respect to the electron and ion mean free paths, and the non-local EDF in the positive column is

$$f_0(w) = \text{const} \times \int_w^{\varepsilon_1} \frac{dW'}{V^3(w')/\nu(w')} \tag{7.65}$$

without elastic losses. This situation is similar to the experimental conditions described in [44]. The key contradiction is that the collisionless ion flux generated in the anode vicinity has a density, which falls off towards the plasma. On the other hand, the electron flux with the EDF (equation (7.65)) without elastic losses gives an electron density falling off towards the anode. So the space charge density which was negative right at the anode rises towards the plasma, and a smooth plasma–sheath matching becomes impossible.

In [48], it is concluded that the phenomenon of the opposite field and the sign of the anode fall in short discharges are interrelated. In this case, two main scenarios may be expected:

(1) At low pressures, when there is one point of the inversion of the sign of the electric field (FR) in the near-cathode plasma of NG and FDS at the maximum of the plasma density, the anode fall is negative and the magnitude of the AF is small, there is no ionization and the anode area is dark.

(2) At increasing pressure, two FR points, a positive AF sign, and the potential fall comparable to the ionization potential of the gas should be expected. Intensive ionization takes place directly at the anode and glows brightly.

In [48], relatively simple probe measurements of the floating potential and optical observations of the anodic glow were proposed and implemented. These measurements enable a quick determination of which of the above scenarios can be implemented in practice. In determining the spatial distribution of the plasma potential, the second derivative of the probe current, the zero of which is identified with the space potential at a given point, is usually measured. However, measuring the second derivative of the probe current is a rather complex experimental task, especially in the cathode region of a discharge.

The measurements in [48] were carried out in a cylindrical glass discharge tube with radius $R_d = 37$ mm with a movable flat molybdenum cathode with a diameter of 56 mm and a flat molybdenum anode with a diameter of 60 mm. A movable cylindrical probe with a length of 3 mm and a diameter of 0.2 mm, which emerges from the center of the anode and allows one to carry out measurements along the axis of the discharge, was located in the tube. The maximum distance between the anode and cathode was 165 mm.

In figure 7.27, the distributions of the space potential along the length of the discharge gap at two typical investigated pressures of nitrogen (20 and 35 Pa, respectively) are shown. Simultaneously with the measurement of the second derivatives, the floating potential of the probe relative to the anode was determined. Figure 7.27 shows that, at a lower gas pressure, the measured potentials are small in absolute value and differ from the anode potential in the range from 1 to 2 V.

The observed pattern corresponds to scenario 1, i.e., a single point of field inversion and a negative anode fall. At the same time, it can be seen that, when the pressure increases, there is a significant rise in the measured potentials to their absolute values relative to the anode, which corresponds to the ionization potential of the gas. This corresponds to scenario 2, i.e. the formation of the second point of

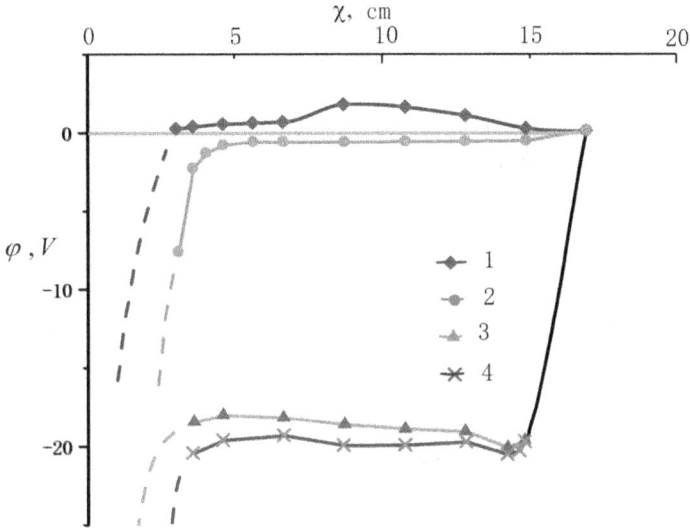

Figure 7.27. Potential distribution of the space (1, 3) and floating potential (2, 4) along the length of the discharge gap (nitrogen, the pressures are 0.2 (1, 2) and 0.35 Pa (3, 4), discharge current is 10 mA), x is the distance from the cathode. (1) Potential of the space, pressure is 20 Pa; (2) floating potential, pressure is 20 Pa; (3) space potential, pressure is 35 Pa; (4) floating potential, pressure is 35 Pa. From [48].

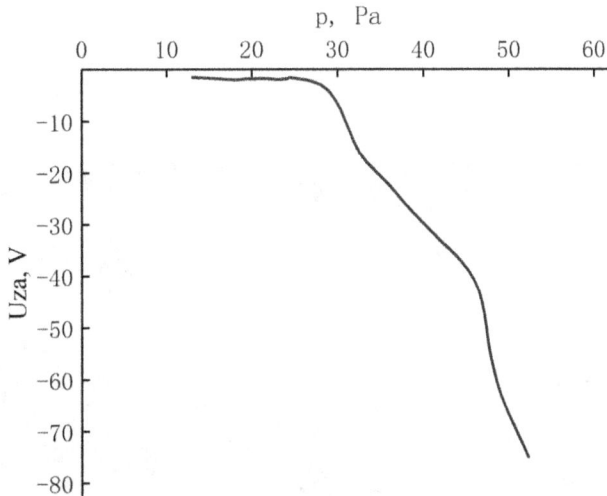

Figure 7.28. Dependence of the floating potential on pressure (nitrogen, discharge current is 10 mA, x = 10 cm to the cathode). After [48].

the field inversion in the plasma and further again in the direct field toward the anode and a positive anode fall. In both cases, the spatial distributions of potential in plasma itself behave similarly, changing little in magnitude and forming weak potential wells for electrons with a depth from 1 to 2 eV. Since the electron temperature T_e in the near-cathode plasma of NG and FDS is low (<1 eV), the difference of potentials, which corresponds to several T_e/e, also has a small value.

A sharp change in the absolute value of the plasma potential relative to the anode in the transition from scenario 1 above to scenario 2 can be seen from the results of measurements of the dependence of the floating potential on the gas pressure for one point inside the gas discharge tube (figure 7.28). It can be seen that, at low nitrogen pressures (from 10 to 29 Pa in this study), the floating potential changes little and remains small (less than 2 V). At the same time, with a pressure increase from 29 to 55 Pa, a sharp increase to 20 V is observed, followed by a more gradual one with gas pressure in absolute value up to 70–80 V. Apparently, this growth is associated with an increase in the length of the direct field area and the potential drop arising there. Figure 7.29 shows the glow discharge for the conditions of figure 7.27. It can be seen that, at low pressures, when $L \leqslant \Lambda_f$ (Λ_f is the characteristic scale of the decay of plasma density in FDS) the anode area is dark and the area of negative glow extends up to the anode. At the same time, when the pressure rises $L > \Lambda_f$, the dark area of the Faraday dark space and glowing film are observed directly near the anode.

Thus, the formation of the potential well for slow electrons on the spatial distribution of the space potential in the near-cathode plasma of NG and FDS can be predicted by the nature of the behavior of the floating potential depending on the experimental conditions. When the floating potential is low in absolute value (in our case, less than 2 V), the anode fall is negative (repulsive electrons). When the floating potential relative to the anode reaches tens of volts in absolute value, it can be stated that the anode fall is positive. In turn, as a result of simple visual observations of the nature of the anode glow, the following conclusions can be drawn.

In the case when the anode is dark, the potential well for electrons up to the anode and a negative anode fall that is small (less than a few volts), the absolute value

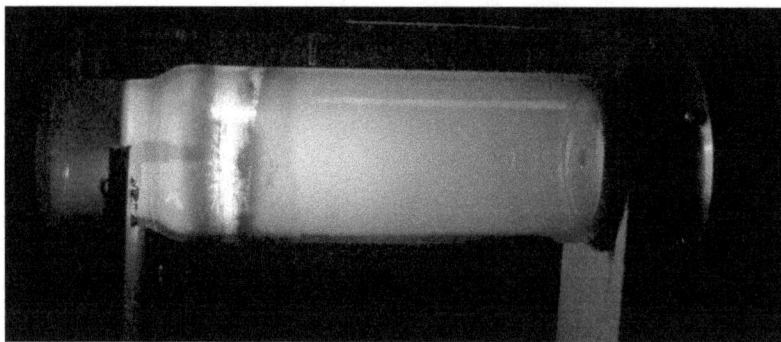

Figure 7.29. Photograph of discharge for conditions of figure 7.27. From [48].

should be expected. At a bright anode glow, a positive anode fall should be expected, the value of which is of the order of or higher than the gas excitation potential. However, in this case, it is technically possible that the direct electrical field along the entire discharge gap exists (from the cathode to anode) when the potential of the plasma changes monotonically and the presence of the potential well is the most probable.

References

[1] Raizer Y P 1991 *Gas Discharge Physics* (Berlin: Springer)

[2] Von Engel A 1955 *Ionized Gases* (Oxford: Clarendon)

[3] Loeb L B 1939 *Fundamental Processes of Electrical Discharge in Gases* (New York: Wiley)

[4] Vedenov V 1982 *Physics of Electro-discharge CO_2 Lasers* (Moscow: Energoatomizdat) [in Russian]

[5] Kolobov V I and Tsendin L D 1992 Analytic model of the cathode region of a short glow discharge in light gases *Phys. Rev.* A **46** 7837

[6] Kudryavtsev A A and Tsendin L D 2000 DC glow discharge *Encyclopedia of Low-Temperature Plasma* vol 2 ed V E Fortov (Moscow: Nauka), p 18 [in Russian]

[7] Druyvesteyn M J and Penning F M 1940 The mechanism of electrical discharges in gases of low pressure *Rev. Mod. Phys.* **12** 87

[8] Surendra M, Graves D B and Jellum G M 1990 Self-consistent model of a direct-current glow discharge: Treatment of fast electrons *Phys. Rev.* A **41** 1112

[9] Goto M and Kondon Y 1998 Monte Carlo simulation of normal and abnormal glow discharge plasmas using the limited weight probability method *Jpn. J. Appl. Phys.* **37** 308

[10] Fiala A, Pichford L C and Boeuf J P 1994 Two-dimensional, hybrid model of low-pressure glow discharges *Phys. Rev.* E **49** 5607

[11] Donko Z, Hartmann P and Kutasi K 2006 On the reliability of low-pressure dc glow discharge modeling *Plasma Sources Sci. Technol.* **15** 178

[12] Kudryavtsev A A, Morin A V and Tsendin L D 2008 Role of nonlocal ionization in formation of the short glow discharge *Tech. Phys.* **53** 1029–40

[13] Boeuf J P 2003 Plasma display panels: physics, recent developments and key issues *J. Phys. D: Appl. Phys* **36** R53

[14] Phelps A V 2001 Abnormal glow discharges in Ar: experiments and models *Plasma Sources Sci. Technol.* **10** 180

[15] Granovsky V L 1971 *Electric Current in Gas: Stable Regime* (Moscow: Nauka) [in Russian]

[16] Ulyanov K N 1972 *Teplofiz. Vys. Temp* **10** 931

[17] Babich L P 2005 Analysis of a new electron-runaway mechanism and record-high runaway-electron currents achieved in dense-gas discharges *Phys. Usp.* **48** 1015

[18] Wilhelm J and Kind W 1965 *Zur Theorie des Glimmlichts einer Niederdruckentladung Beitr. Plasmaphys* **5** 395–403

[19] Peres I, Quadoudi N, Pichford L C and Boeuf J P 1992 Analytical formulation of ionization source term for discharge models in argon, helium, nitrogen, and silane *J. Appl. Phys.* **72** 4533

[20] Rozsa K, Gallagher A and Donko Z 1995 Excitation of Ar lines in the cathode region of a dc discharge *Phys. Rev.* E **52** 913

[21] Maric D, Kutasi K, Malovic G, Donkó Z and Petrovićet Z L 2002 Axial emission profiles and apparent secondary electron yield in abnormal glow discharges in argon *Eur. Phys. J.* **21** 73

[22] Maric D, Hartmann P, Malovic G, Donkó Z and Petrović Z L 2003 Measurements and modeling of axial emission profiles in abnormal glow discharges in argon: heavy-particle processes *Phys. D: Appl. Phys.* **36** 2639

[23] Boeuf J P and Pichford L C 1995 Field reversal in the negative glow of a DC glow discharge *J. Phys. D: Appl. Phys* **28** 2083

[24] Gottscho A, Mitchell A, Scheller G R, Chan Y Y and Graves D B 1989 Electric field reversals in dc negative glow discharges *Phys. Rev.* A **40** 6407

[25] Kudryavtsev A A and Toinova N A 2005 A criterion for electric field reversal in the negative glow region of a short DC glow discharge *Tech. Phys. Lett.* **31** 370

[26] Phelps A V and Petrovic Z L 1999 Cold-cathode discharges and breakdown in argon: surface and gas phase production of secondary electrons *Plasma Sources Sci. Technol.* **8** R21

[27] Phelps A V, Pichford L C, Pedoussat C and Donko Z 1999 Use of secondary-electron yields determined from breakdown data in cathode-fall models for Ar *Plasma Source Sci. Technol.* **8** B1

[28] Apostol I, Kagan Y M and Lyagushchenko R I *et al* 1976 Sov *Phys. Tech. Phys.* **21** 1168

[29] Kudryavtsev A A and Tsendin L D 2002 Townsend discharge instability on the right-hand branch of the Paschen curve *Tech. Phys. Lett.* **28** 1036

[30] Tsendin L D 1995 Electron kinetics in non-uniform glow discharge plasmas *Plasma Sources Sci. Technol.* **4** 200

[31] Velikhov E P, Golubev V S and Pashkin S V 1982 Glow discharge in a gas flow *Sov. Phys. Usp.* **25** 340–58

[32] Rozhansky V A and Tsendin L D 2001 *Transport Phenomena in Partially Ionized Plasma* (London: Taylor & Francis)

[33] Dmitriev A P, Rozhansky V A and Tsendin L D 1985 Diffusion shocks in an inhomogeneous current-carrying collisional plasma *Sov. Phys. Uspekhi* **28** 467–83

[34] Demidov V I, DeJoseph C A and Kudryavtsev A A 2006 Nonlocal effects in a bounded afterglow plasma with fast electrons *IEEE Trans. Plasma Sci.* **34** 825

[35] Demidov V I, DeJoseph C A and Kudryavtsev A A 2005 Anomalously high near-wall sheath potential drop in a plasma with nonlocal fast electrons *Phys. Rev. Lett.* **95** 215002

[36] Bogdanov E A, Adams S F, Demidov V I, Kudryavtsev A A and Williamson J M 2010 Influence of the transverse dimension on the structure and properties of dc glow discharges *Phys. Plasmas* **17** 103502

[37] Kudryavtsev A A, Morin A V and Tsendin L D 2008 Role of nonlocal ionization in formation of the short glow discharge *Tech. Physics* **53** 1029

[38] Adams S F, Demidov V I, Bogdanov E, Koepke M E, Kudryavtsev A A and Kurlyandskaya I P 2016 Control of plasma properties in a short direct-current glow discharge with active boundaries *Phys. Plasmas* **23** 024501

[39] Demidov V I, Kudryavtsev A A, Kurlyandskaya I P and Stepanova O M 2014 Nonlocal control of electron temperature in short direct current glow discharge plasma *Phys. Plasmas* **21** 094501

[40] Bogdanov E A, Demidov V I, Kudryavtsev A A and Saifutdinov A I 2005 Is the negative glow plasma of a direct current glow discharge negatively charged? *Phys. Plasmas* **22** 024501

[41] Lieberman M and Lichtenberg A 2005 *Principles of Plasma Discharges and Materials Processing* (New York: Wiley)

[42] Compton T and Langmuir I 1930 Electrical discharges in gases. part I. survey of fundamental processes *Rev. Mod. Phys.* **2** 123
Compton T and Langmuir I 1931 Electrical discharges in gases part II. Fundamental phenomena in electrical discharges *Rev. Mod. Phys.* **3** 191

[43] Von Engel A 1941 A theory of the anode fall in glow discharges *Philos. Mag.* **32** 417

[44] Klarfeld B N and Neretina N A 1958 Anode region in gas discharge at low pressure. Part I. Effect of anode shape on sign and magnitude of anode drop Sov. Phys *Tech. Phys.* **3** 1960
Klarfeld B N and Neretina N A 1958 Anode region in a low-pressure gas discharge. Part III. Production of additional plasma at the anode *Sov. Phys. Tech. Phys.* **5** 169

[45] Akishev Y S, Dvurechenskii S V, Napartovich A P, Pashkin S V and Trushkin N I 1982 Study of the plasma-column and anode region of an axial discharge in nitrogen and air *High. Temp.* **20** 27–34

[46] Tsendin L D 1986 Anode region of a glow discharge *Sov. Phys. Tech. Phys.* **56** 279

[47] Tsendin L D 2011 On the magnitude and sign of the anode potential fall in a gas discharge *Tech. Phys.* **56** 1693

[48] Prokhorova E I, Kudryavtsev A A, Platonov A A and Slyshov A G 2017 Correlation between reversion of signs of the electric field in the near-cathode plasma and anode fall potential in a short DC glow discharge *Tech. Phys.* **62** 1122

Chapter 8

Positive column of dc glow

Among the numerous glow discharge manifestations the dc positive column (PC) was, maybe, the most familiar, and what is more important, the most thoroughly investigated plasma object. For more than a century of PC research, immense amounts of information were stored. The simplicity of its realization, simple one-dimensional geometry, its stationarity, the relative cheapness of its experimental investigation and diagnostics have made the PC a traditional benchmark and testing field for many novel ideas and diagnostics in plasma physics.

8.1 Main properties of dc glow positive column

The PC connects the cathode and the anode regions of the discharge; it represents the conducting media, which transports current between the cathode and the anode [1]. If a discharge tube is long enough, the PC occupies its main part; only the PC length L_{PC} goes up, when the discharge gap L becomes longer. The potential profile in the electrode regions doesn't depend on L_{PC}; the discharge voltage varies only due to the L_{PC} variation. Only when the gap L is so short that there is no place left for the PC and it disappears, do the electrode regions start to overlap and essential discharge rearrangement occur (see chapter 7). It is said that in spite of the fact that the PC often represents the longest discharge part, generally, it is unnecessary for discharge maintenance. In other words, the PC represents an autonomous self-replicating system.

The PC properties are practically independent of the electrode characteristics (for example, material, size, form and temperature) and on the discharge volume form (twisted or straight). The classic PC, which has been investigated for more than a century, operates in cylindric dielectric tubes (typically made from glass or quartz) and its radius usually ranges from less than one millimeter (capillary discharge) up to several centimeters. The tubes are filled with various gases, metal vapor or their mixtures and typical pressure values are from mTorrs up to atmospheric pressure and higher. Discharge current may be in a large current range from micro-amperes up to tens of amperes.

The PC is widely used in the lighting industry, as an active media of gas and plasma lasers, in plasma-chemistry, and in spectral analysis. It is also widely used in various branches of modern technology, such as surface treatment and coating, sputtering, thin film formation, etc.

The electric potential profile in a stationary longitudinally-uniform PC is equal to

$$\Phi(r, x) = -E_x x + \varphi(r). \tag{8.1}$$

The longitudinal field E_x maintains maintains the plasma generation to balance the charged particle loss. The Joule electron heating in this field forms the EDF and the longitudinal (mainly electron) current. The radial ambipolar field $E(r) = -\nabla\varphi(r)$ is responsible for maintenance of plasma quasi-neutrality, by forming equal electron and ion radial fluxes and accumulating the surface charge. At low current and/or pressure, the charged particle removal is controlled by the radial ambipolar diffusion with subsequent recombination on the tube wall. As current/pressure rises, the contribution of the volumetric recombination to the removal of charge particles increases. The observed dependences of the reduced electric field E_x/p (which is proportional to the electron energy gained on the mean free path) on pR_d, which represents the R/λ ratio, for the PC in several noble and molecular gases are shown in figure 8.1 [1]. The plasma density, as well as the current density j and plasma luminosity, are usually maximal at the tube axis and go down towards the tube wall. The radial field equals zero at the axis and goes up towards the wall, reaching its maximal values in the wall-adjacent space-charge sheath. The voltage between the tube axis and wall $\Phi_w = \Phi_{pl} + \Phi_{sh}$, where Φ_{pl} is the voltage over the quasi-neutral plasma and Φ_{sh} is the voltage over the sheath. It suppresses the radial electron diffusion. Here, the value of Φ_w is of the order of T_e/e. As examples, in figure 8.2 [1, 2] the radial profiles of plasma potential in PC of mercury are shown, which are typical for the low and medium-pressure discharges.

The first PC theories were proposed by Langmuir and Tonks [3] for the free-flight or free-fall ($R_d \ll \lambda_i$) regime and by Schottky [4] for the diffusive case, when the

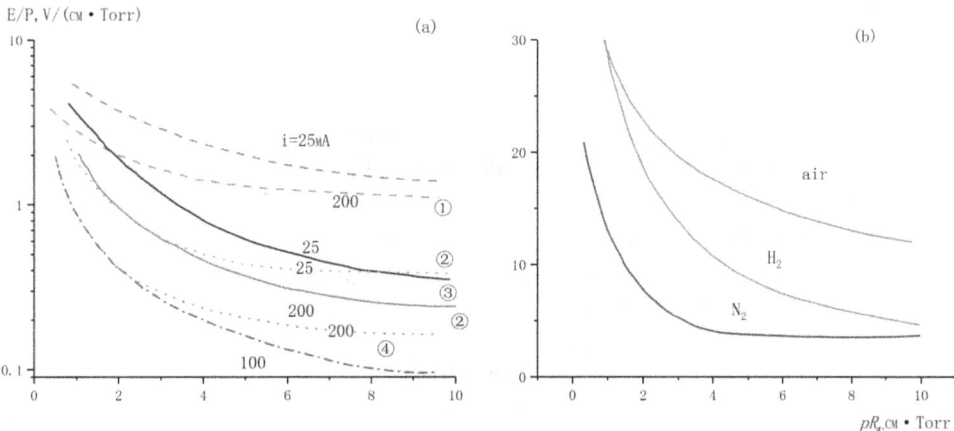

Figure 8.1. The observed electric field E_x/p in noble (a) and molecular (b) gases in PC dependent on pR_d. In (a) gases: He (1), Ar (2), Ne (3) and Xe (4); the numbers near the curves are the current values in mA. After [2].

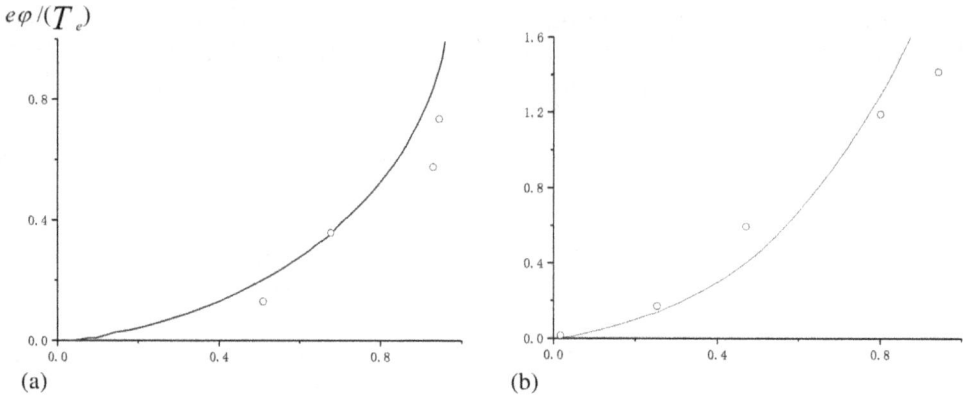

Figure 8.2. The potential profile in the PC of mercury vapor for the free-fall case $pR_d = 3 \times 10^{-4}$ Torr cm (a) and for the diffusion-dominated ion motion $pR_d = 3 \times 10^{-2}$ Torr cm (b). Solid lines are calculations with Langmuir–Tonks theory (a) and Schottky theory (b). Points are experiment. After [2].

radial ion motion is controlled by diffusion ($R_d \gg \lambda_i$). In both cases the stationary condition of the PC maintenance reduces to

$$\nu_i \tau_s = 1. \tag{8.2}$$

In the free-flight regime the quasi-neutral equation for a cylindrical PC is [2, 5]

$$\exp\left(\frac{e\varphi(r)}{T_e}\right) = \frac{1}{r} \int_0^x \frac{r'\nu_i \exp\left(\dfrac{e\varphi(r')}{T_e}\right) dr'}{\sqrt{2e(\varphi(r') - \varphi(r))/M}}. \tag{8.3}$$

Its solution [3] has the electric field singularity, which coincides with the plasma–sheath interface [5]. The corresponding voltage at the plasma boundary is

$$\Phi_{pl} = \frac{1.155 T_e}{e}. \tag{8.4}$$

Neglecting the sheath thickness with respect to R_d, if follows from equation (8.3) that (8.2) will be

$$\nu_i = \frac{0.772\sqrt{\dfrac{2T_e}{M}}}{R_d}. \tag{8.5}$$

So the average ion velocity at the plasma–sheath interface is $u_i = 1.14\sqrt{T_e/M}$, which practically coincides with the Bohm criterion. The average plasma lifetime is $\tau_s = R_d/\sqrt{1.19 T_e/M}$.

In the diffusive case the solution is given by [4]

$$n(r) = n(0) J_0\left(\frac{2.405 r}{R_d}\right); \qquad \nu_i = \frac{2.405^2 D_a}{R_d^2}. \tag{8.6}$$

The plasma voltage in the diffusion-dominated case is

$$\Phi_{pl} = \frac{T_e}{e} \ln\left(\frac{R_d}{\lambda_i}\right). \tag{8.7}$$

The potential profiles corresponding to equations (8.3) and (8.6) are shown by the full lines in figure 8.2. The sheath voltage in the case of the collisionless sheath is

$$\Phi_{sh} \approx \frac{kT_e}{e} \ln \sqrt{\frac{M}{m}} . \tag{8.8}$$

Thus, in equation (8.2) the average plasma lifetime in the Langmuir-Tonks theory is expressed via the Bohm speed (8.5) and in the Schottky theory is expressed via time of the ambipolar diffusion (8.6). There are interpolation formulas combining both theory. For estimation, the following formula can be used, $\frac{1}{\tau_s} = D_a/\left(\frac{R_d}{2.405^2}\right) + \frac{\sqrt{\frac{1.19T_e}{M}}}{R_d})$. A more accurate approximation for τ_s was given in [6]

$$\frac{1}{\tau_s} = D_s/\left(\frac{R_d}{2.405}\right)^2, \tag{8.9}$$

where for the effective coefficient D_s one can use an interpolation

$$D_s = D_a/\left(1 + \frac{2\nu_i}{\nu_{ia}}\right) \tag{8.10}$$

with $D_a = D_i(1 + T_e/T_0)$ being the ambipolar diffusion coefficient and ν_{ia} the ion-neutral collision frequency. Relationship (8.2) defines the ionization frequency ν_{ia} in terms of the PC external parameter pR_d. For the Maxwellian EDF the frequency ν_i depends on the electron temperature T_e. If the energy dependence of the ionization cross-section is linear, $\sigma_i(w) = \sigma_{i0}(w/\varepsilon_i - 1)$, we have

$$\nu_i(T_e) = \sqrt{\frac{2T_e}{m}} N\sigma_{i0}\left(1 + \frac{2T_e}{\varepsilon_i}\right)\exp\left(-\frac{\varepsilon_i}{T_e}\right). \tag{8.11}$$

So equation (8.2) is of the form of the transcendent equation for T_e, which is of the order of several eV. Thus equations (8.2), (8.7) and (8.11), complemented by the equation for the (electron) current

$$j \approx en_e u_e = en_e b_e E, \tag{8.12}$$

allow the calculation of all plasma parameters if the cross-section set is given.

In the electron energy balance, the energy is supplied by the Joule heating in the field E_x. At low pressure the energy loss is controlled by the inelastic electron-neutral collisions, which result in excitation of the neutrals. As usual this energy is quickly emitted by radiation and the temperature T_0 of the neutrals remains low. The temperature T_0 is controlled by the thermal conductivity to the tube wall and is maintained close to room temperature. On the other hand, as the plasma is generated by the electron impact ionization, the electron energies are to be of the order of ε_i, i.e., the EDF energy scale is to be of the order of a few eV. So the electron energies in the PC are one or two orders of magnitude higher than T_0.

In figure 8.3 the neutral gas temperature at the PC axis in noble gases is presented dependent on the discharge current. As in molecular gases a considerable fraction of the energy, transferred from electrons to excitation of the rotational and vibrational

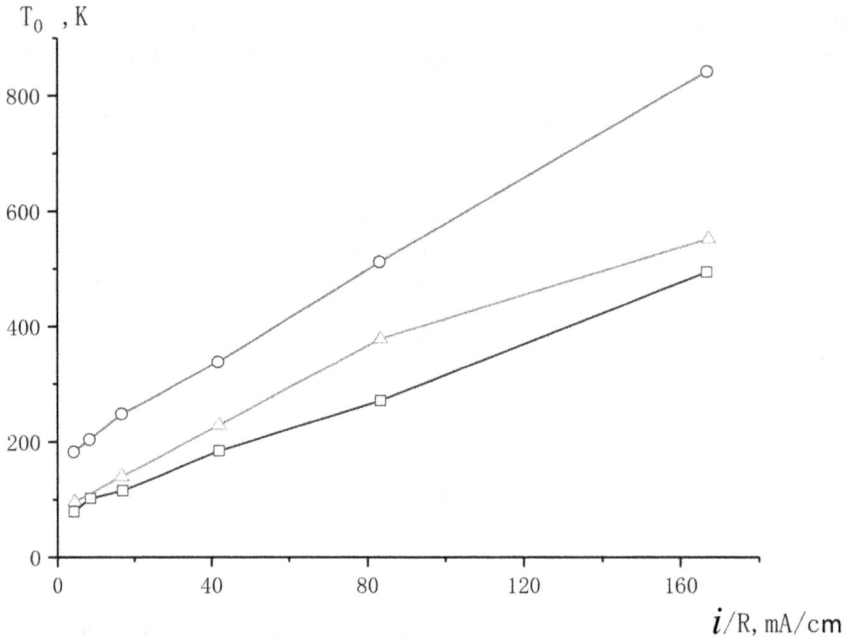

Figure 8.3. The current dependence of the gas temperature T_0 (K) at the tube axis for $pR_d = 24$ Torr cm in noble gases: argon (red), helium (green) and neon (blue). From [2].

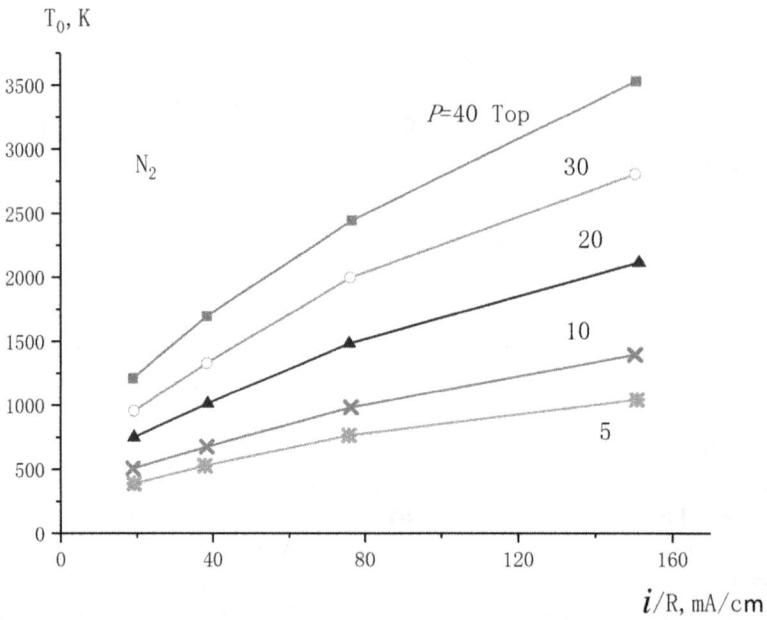

Figure 8.4. The current dependence of the gas temperature T_0 (K) at tube axis for various p and $R_d = 1$ cm at the tube axis in nitrogen. From [2].

molecular energy levels, is transformed afterwards into kinetic energy of molecules. So the neutral gas temperature in nitrogen, shown in figure 8.4, is considerably higher, than in atomic gases. The gas temperature radial profiles in the noble gases PC are presented in figure 8.5.

The weak point of the standard traditional PC theories consisted of the arbitrary assumption of the Maxwellian EDF. It allowed the closing of the equation set and all the discharge characteristics to be expressed in terms of the external parameters. However, for that assumption there is no physical reason. The EDFs in the PC can strongly deviate from the Maxwellian ones. For the EDF Maxwellization due to electron–electron collisions, an extremely high ionization degree (exceeding 10^{-2}) is necessary, which can hardly be reached in the low-pressure glows of interest.

The pressure and current rise leads to the energy fraction increasing, transferring from electrons to neutrals in elastic collisions and to decreasing the neutral thermal conductivity to the tube wall. As a result, the electron and the neutral energies converge and plasma could approach the local thermodynamic equilibrium (LTE). Such a situation prevails, for example, in a PC of a high-pressure arc. As was already mentioned in the introduction the great distinction of the characteristic energy scales

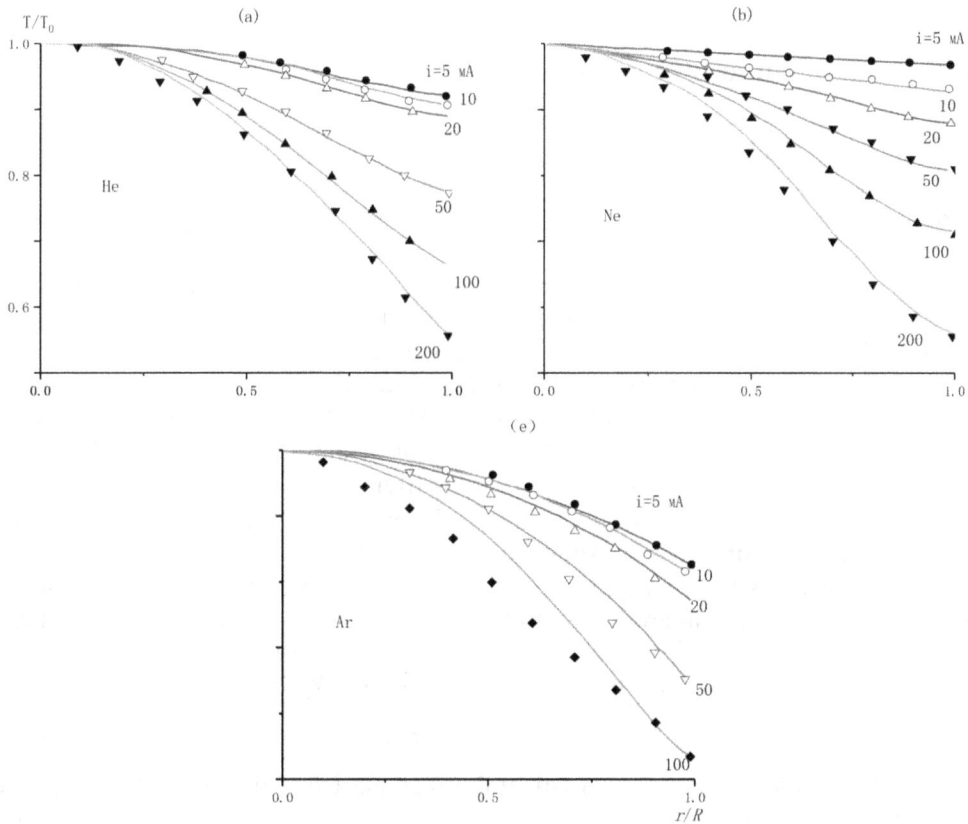

Figure 8.5. The radial profiles of the gas temperature in noble gases at $pR_d = 24$ Torr cm. The values of T_0 are from figure 8.4. From [2].

between electrons and the neutral particles, which is typical not only for PC, but for low- and medium-pressure glows, is the main factor which makes these phenomena extremely far from local thermodynamic equilibrium (LTE) and demanding of far more detailed (and complicated) kinetic treatment. This non-equilibrium manifests itself not only in the necessity of kinetic analysis of the electron component, but also in the fact that practically every process is not balanced by the inverse. For example, charged particle generation in collisions in the plasma bulk is balanced by their recombination on the cold surface of the tube wall. Even if the recombination in the plasma volume dominates, its channels (dissociative or collisional-radiative) are different from the inverse to the usually dominating ionization by electron impact. Accordingly, the ionization degree is several orders of magnitude higher, than given by the Saha formula with the neutral temperature T_0, and by several orders lower, than one given by the Saha expression with the electron temperature T_e. As the typical plasmas in glows is optically thin (with the exception sometimes of central parts of the resonant lines), free escape of the radiation is possible. So the electron impact excitation processes are balanced not by the super-elastic collisions, but mainly by the radiation escape. So the population of the excitation levels has nothing in common with the Boltzmann distribution, and the radiation density does not obey the Planck formula.

8.2 Positive column in atomic gases

For full self-consistent PC description it is necessary to express its internal characteristics, such as EDF, both radial and longitudinal electric fields, radial profiles of the plasma components densities and of the densities of the excited particles and of the ionization rate, in terms of the external parameters: current density, gas pressure, tube radius and collision cross-sections.

There exist considerable differences between discharges at low and high pressures in respect of the electron kinetics. One of the most characteristic properties of low-pressure discharges is that they are surprisingly quiet and homogeneous. In this case the EDF relaxation length exceeds the tube radius, the radial diffusive motion of electrons produces the dominant effect on EDF, the EDF factorization is impossible, and here its form is determined by the whole radial potential profile (the non-local EDF). In the atomic gases the EDF relaxation length is usually of the order of the energy relaxation length in elastic collisions $\lambda_\varepsilon \sim \sqrt{M/m}\,\lambda \sim 10^2\lambda$. So the criterion $R_d \sim \lambda_\varepsilon$ corresponds, roughly speaking, to $pR_d \sim 10$ Torr cm. As in the molecular gases the energy loss in the quasi-elastic collisions (excitation of the rotational and vibrational levels) is significant, the transition from the local and non-local EDF formation regimes lies considerably lower, at $pR_d \sim 1$ Torr cm.

It is to be noted that from the non-locality criterion $R_d < \lambda_\varepsilon = T_e/(eE_z)$ it follows

$$E_r \sim T_e/R_d > E_x,$$

i.e., in the non-local case the radial ambipolar electric field always exceeds the heating longitudinal field.

Contrary to low pressure, discharges at high pressures are subjected to various pinching (pinching is the major factor complicating numerous practical applications). The reason is that in the local approximation, which is valid at high pressures, the EDF tail and, consequently, the ionization rate are rigidly bound to a local electric field which heats up the electrons. This being so, any process strengthening the field in plasma contractions and thereby the ionization rate in them leads to ionization instability. As a result of its development, sharply non-uniform plasma density profiles are formed. The non-uniform Joule heating of neutral gas and Maxwellization of the EDF tail caused by inter-electron collisions are indicated as the main instability mechanisms.

At high pressures, when the EDF relaxation length λ_ε is shorter than the tube radius, the electrons are accelerated in the longitudinal electric field without noticeable diffusive displacement in radial direction. This permits the radial EDF gradient in the kinetic equation (local approximation for EDF) to be neglected. The local EDF $f_0(r,z,t,w)$ can be factorized into the product of the density $n(r,z,t)$ by the function $F_0(w)$, which parametrically depends on the field E/p. The exception represents the region near the tube walls, the length of which is about the EDF relaxation length λ_ε.

Historically, the local approach for the definition of EDF is often used as the description of a positive column and the analysis of physical processes even when it is obviously inapplicable. So it is interesting to compare the predictions of the local approximation and more general (non-local) theory to the correct simulations and calculations. The majority of such comparisons have been used for noble gases using the same cross-section set, for example, in [8–11]. Further, we will consider the basic results presented in [10, 11]. Here, we only briefly recall the features of the model important for the PC [11]. The model is based on a fluid description of ions and neutral species (ground state or excited) using a drift–diffusion approximation for the particle flux. The continuity equations are solved for the mass density of each ion and neutral components of the plasma. The transport coefficients (mobility μ_e and diffusion D_e) and rates of electron-induced chemical reactions S_e are calculated using the electron energy distribution function (EDF) f_0 obtained as a solution of the electron Boltzmann equation. We recall that the EDF in the calculation of which terms with spatial gradients in the kinetic equation play an important role is here referred to as non-local. In the non-local case for the PC analysis it is more convenient to define the total energy ε as the sum of the kinetic w, and of the potential energy in the radial field $e\varphi(r)$, including the uniform field E_z into the diffusion coefficient along energy D_ε.

For comparison, the kinetic equation in the conventional local approximation could also have been solved, in which the radial electric field and the radial gradients in the kinetic equation were ignored. The traditional fluid model could be also used, when the parameters of the electron gas are found using the fluid equations for the balance of the electron density and energy [14]. The self-consistent electric field is found from Poisson's equation.

In [10, 11] self-consistent simulations of a PC in atomic gases in a tube of radius $R_d = 1$ cm at $p = 0.1$–10 Torr were performed. The most frequently employed and accepted three-level model of an argon atom with one metastable state (index m) was used, taking into account the eight main reactions listed in table 8.1.

Table 8.1. The set of reactions used for the three-level scheme of terms of an argon atom. From [10].

No.	Reaction	$\Delta\varepsilon$, eV	Constant	Commentary
1	$e + Ar \longrightarrow e + Ar$	–	Cross-section	Elastic scattering (momentum transfer)
2	$e + Ar \longrightarrow e + Ar_m^*$	11.55	Cross-section	Excitation and de-excitation of the metastable level
3	$e + Ar \longrightarrow 2e + Ar^+$	15.9	Cross-section	Direct-ionization from the ground state
4	$e + Ar_m^* \longrightarrow 2e + Ar^+$	4.35	Cross-section	Stepwise ionization from the metastable level
5	$e + Ar_m^* \longrightarrow e + Ar_r^*$	0.07	$k_q = 2 \times 10^{-13}$ m^3 s^{-1}	Quenching of the metastable level via the transition to the resonance level (11.67 eV)
6	$e + Ar \longrightarrow e + Ar$	11.5	Cross-section	Total excitation by electron impact
7	$Ar_r^* \longrightarrow Ar + \hbar\nu$	–	$A_R = 10^6$ s^{-1}	Resonance emission with allowance for self-absorption ($\lambda = 106.4$ nm)
8	$Ar_m^* \longrightarrow \begin{cases} e + Ar^+ + Ar \\ e + Ar_2^+ + Ar \end{cases}$	–	$k_p = 6.2 \times 10^{+16}$ m^3 s^{-1}	Penning ionization

The constants of the processes with the participation of electrons were calculated by convoluting the corresponding cross-sections with the calculated EDF. It can be seen from figure 8.6 (right) that, even at a relatively high pressure, the EDF $f^0(w,r)$ is not factorized in the form

$$f_0(w, r) = n_e(r) \times F_0(w) \tag{8.13}$$

the normalized EDF plotted as a function of the kinetic energy is different at different radii r; i.e., it is non-local.

At the same time, the EDFs $f_0(\epsilon,r)$ plotted as functions of total energy (figure 8.6 (left)) at different radii coincide at $\epsilon < \epsilon^*$ (where ϵ^* is the threshold energy for inelastic processes), without any shift related to normalization and the spatial dependence of the potential; however, they differ in the inelastic energy range $\epsilon > \epsilon^*$ (figure 8.7). It is well known that, in practice, the fact that the EDF $f_0(\epsilon,r)$ of trapped electrons with $\epsilon \leqslant e\Phi_w$ (where Φ_w is the wall potential) does not depend on the radius clearly indicates the non-local character of the EDF (in contrast, the EDF of transit electrons with $\epsilon > e\Phi_w$ depends on the radius). At first glance, it would seem that, in this situation, the EDF consists of a non-local component at $\varepsilon < \varepsilon^*$ and a local component at $\varepsilon > \varepsilon^*$ (a small local 'tail' is attached to the non-local 'body'). However, this is not the case because the fast component at $\varepsilon > \varepsilon^*$ is not factorized in form of equation (8.13). As was noted above, the use of the local approximation for determining the EDF is restricted to the case where this function is factorized in the form of equation (8.13). When the terms with spatial gradients in the kinetic

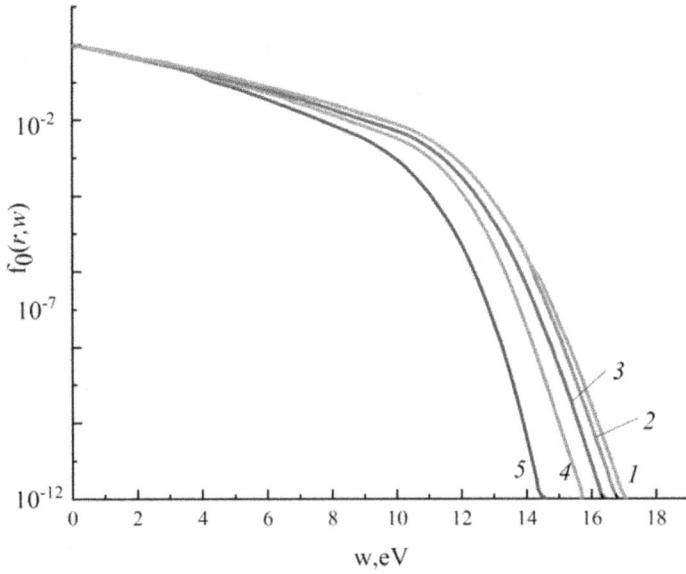

Figure 8.6. Normalized EDFs for $pR_d = 5.4$ cm Torr as functions of kinetic energy: $r = $ (1) 0, (2) $0.2R_d$, (3) $0.4R_d$, (4) $0:6R_d$, and (5) $0:8R_d$. From [11].

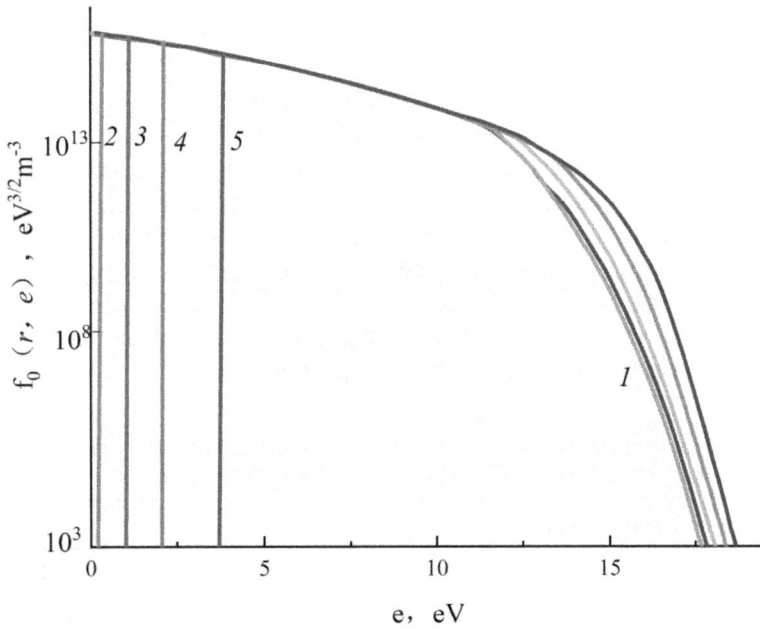

Figure 8.7. EDFs over the total energy for $pR_d = 5.4$ cm Torr. Curves 1–5 correspond to the same radii as in figure 8.6. From [11].

equation cannot be omitted, the EDF is considered as non-local. Depending on the total energy w, the components of this non-local EDF may be either dependent on or independent of the radius.

It follows from table 8.2, which presents the parameter values at the discharge axis at $p = 6$ Torr and $R_d = 1$ cm, and from figures 8.8–8.11, which show the corresponding radial profiles, that the parameters calculated in the local and non-local models differ significantly at medium pressures. The profiles of the radial field in figure 8.9 show that this field exceeds the axial field in most of the PC. This fact indicates that the local approximation is inapplicable for calculating the EDF [15]. This is also confirmed by a decrease in the mean electron energy with radius (figure 8.10). We also note that the electron density profiles can be narrower or broader than the conventional (Bessel) profiles calculated in the fluid model. The reason is a competition between the two effects [11]. First, the ionization is non-uniform over the cross-section, and the density is maximum at the tube axis (see figure 8.13). Second, the average electron density decreases toward the periphery (figure 8.10), which results in a lower coefficient of ambipolar diffusion there. At low pressures, the first effect is dominant and the electron density profiles decrease almost linearly with radius [11]. In the case under consideration, the second mechanism turns out to be more efficient, which results in the opposite effect—the broadening of the profiles in comparison with the local model (figure 8.11). Calculations [10] have shown that the peak of the excitation rate profile shifts from the axis of a discharge toward the periphery due to the non-local character of the EDF when the pressure is increased from low ($pR_d < 1$ cm Torr) to medium ($1 < pR_d < 10$ cm Torr) values.

Simulations [10, 11] revealed an interesting (but little known) effect consisting of the replication of the shape of the body of the EDF in its fast component due to impacts of the second kind (super-elastic collisions) of slow electrons with meta-stable atoms in the reaction $Ar^* + e \rightarrow Ar + \vec{e}$. These processes substantially influence the calculated values of the constants for the excitation reactions with high threshold energies and, accordingly, the densities of highly excited states. Figure 8.12 shows the EDFs calculated for $p = 1$ Torr, $R_d = 1$ cm, and $I = 10$ mA with (curve 1) and without (curve 2) allowance for impacts of the second kind. One can see that a gently sloping pedestal f_{0h}, replicating the shape of the slow

Table 8.2. Comparison of the results of simulations of the plasma parameters at the discharge axis with calculations in the local model for $p = 6$ Torr, $R_d = 1$ cm, and $I = 3$ mA. From [10].

	Local approximation	Full simulation
n_e, cm^{-3}	2.9×10^{10}	1.5×10^{10}
$2\langle\varepsilon\rangle/3$, eV	3.3	3.5
Φ_W, V	68.5	18.5
n_m, cm^{-3}	1.7×10^{11}	5.4×10^{10}
ν_m, s^{-1}	4.4×10^4	1.2×10^4
ν_{ex}, s^{-1}	1.4×10^5	3.1×10^5

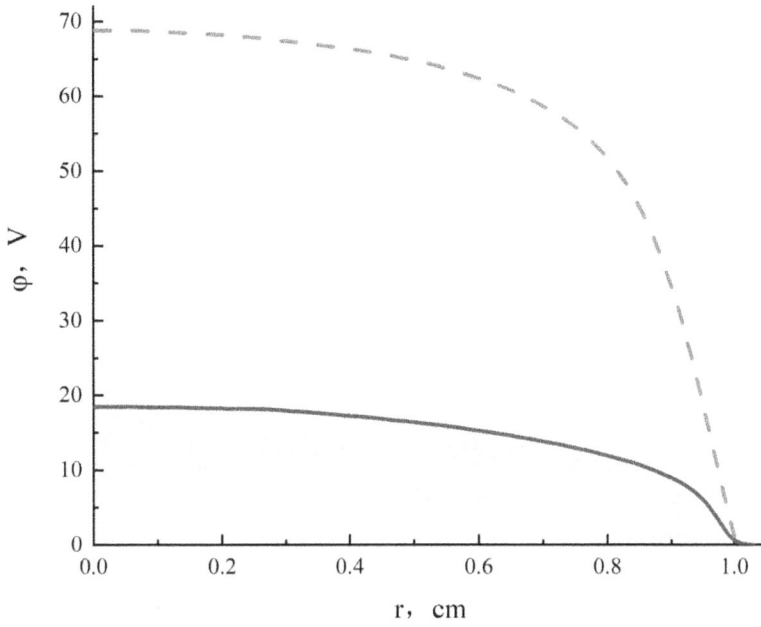

Figure 8.8. Radial profile of the electric potential for $p = 6$ Torr and $I = 3$ mA. The dotted line corresponds to the local approximation. From [12].

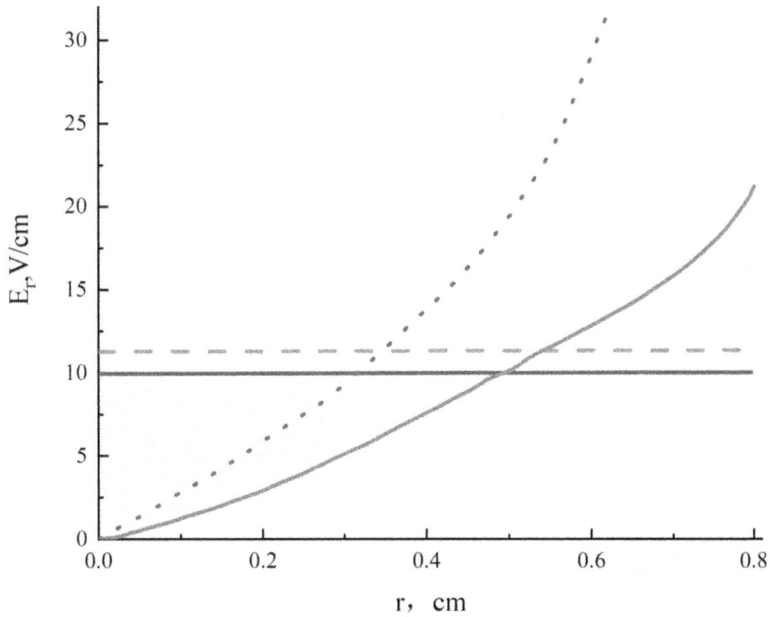

Figure 8.9. Radial profile of the electric field for $p = 6$ Torr and $I = 3$ mA. The dotted lines correspond to the local approximation. The horizontal lines show the corresponding values of the longitudinal electric field E. From [12].

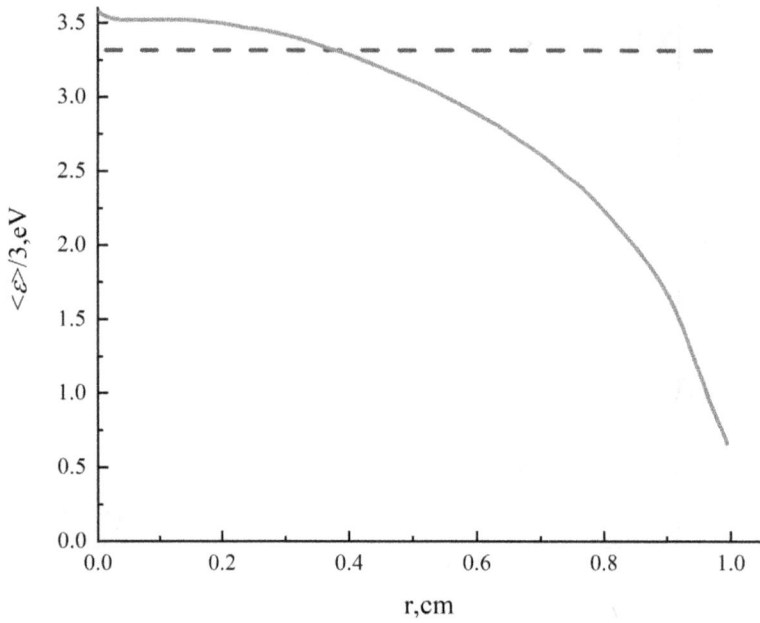

Figure 8.10. Radial profile of the mean electron energy for $p = 6$ Torr and $I = 3$ mA. The dotted line corresponds to the local approximation. From [12].

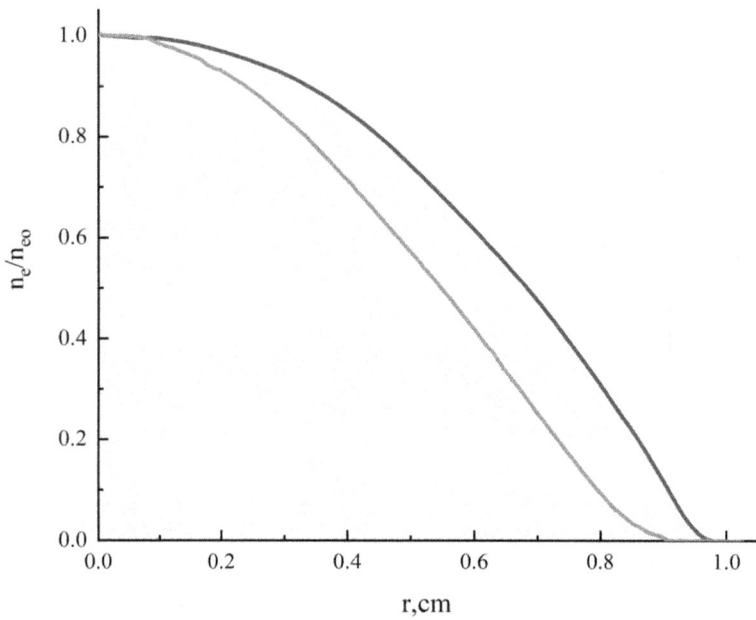

Figure 8.11. Radial electron-density profile for $p = 6$ Torr and 3 mA. The dotted line corresponds to the local approximation. From [12].

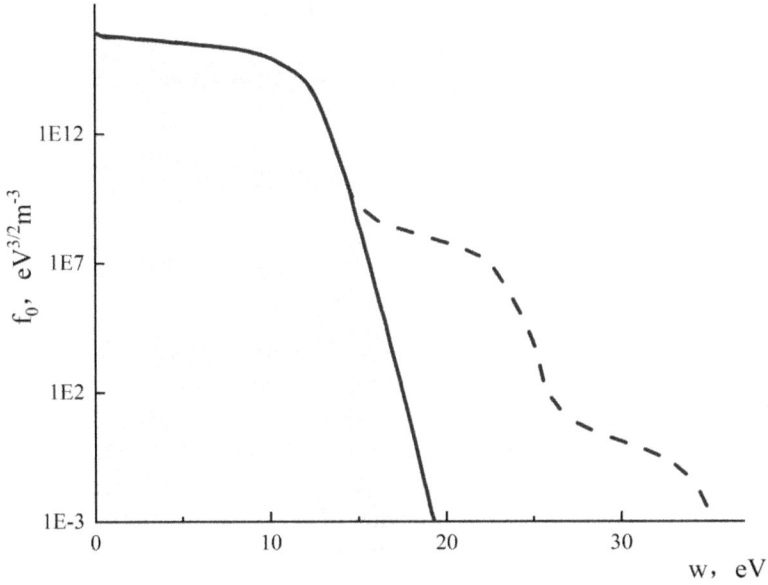

Figure 8.12. EDFs at the axis of a discharge ($r = 0$), calculated (1) with and (2) without allowance for impacts of the second kind for $p = 1$ Torr, $R_d = 1$ cm, and $I = 10$ mA. From [11].

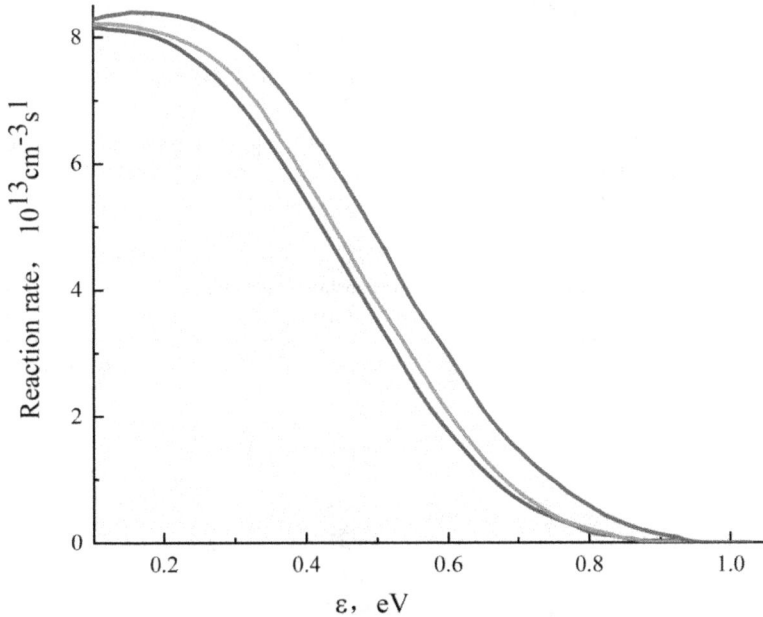

Figure 8.13. Radial profiles of the direct and step-ionization rates (the corresponding curves coincide). The dotted curve shows the density profile of high-energy ($w \geqslant 16 \text{eV}$) electrons, and the dashed line shows the $n_m(r)$ $n_e(r)$ profile (in arbitrary units). The step-ionization rate is plotted in absolute units, whereas the direct ionization rate is multiplied by 780 to bring the plots into coincidence. From [12].

component (the body) of the EDF is added to the usual EDF f_{0t}, which is sharply decreasing above the threshold excitation energy ($\varepsilon^* \simeq 11.55 eV$ for argon). As a result, the fast component of the EDF can be represented as the sum $f_0 = f_{0t} + f_{0h}$. This shape of the EDF can be explained by the fact that the kinetic equation includes the source of fast electrons that is associated with impacts of the second kind and is proportional to the low-energy component (body) of the EDF, $f(w - \varepsilon^*)$ (dotted curve on figure 8.12)

$$f_{0h}(\varepsilon) \simeq \frac{n_m g_a}{N_0 g m} f_0(\varepsilon - \varepsilon^*). \tag{8.14}$$

Roughly speaking, a partial solution taking into account inhomogeneity f_{0h} is added to the rapidly decaying solution of the homogeneous kinetic equation for f_{0t} derived without allowance for this source. Although the amplitude (absolute value) of the additional fast component is lower by a factor of $\sim n_m/N_0$ than the low-energy component of the EDF at $\varepsilon < \varepsilon^*$, it exceeds f_{0h} at energies of several eV above the threshold energy ε^*. This is because the value of T_{et} is small and f_{0t} sharply decreases with increasing energy. Since $f_0(\varepsilon - \varepsilon^*)$ near the threshold energy ε^* varies slightly the high-energy component of the EDF increases substantially as a result of the replication of the body of the EDF into this energy range. In turn, this results in a change in the constants for direct ionization and the excitation for high-lying levels. As an illustration, table 8.3 show the relative change in the constants for direct ionization and the excitation of high-lying excited states when the EDF is calculated with and without allowance for impacts of the second kind of slow electrons with metastable atoms for $N_0 = 10^{17}$ cm^{-3}, $R_d = 1$ cm, and $I = 10$ mA. It can be seen from the table that the results differ markedly; hence, this effect should be taken into account in calculating high-lying excited states in gases that have long-lived metastable states. Since f_{0h} is proportional to n_m/N_0 and its energy dependence is close to that of the EDF of slow electrons ($\varepsilon < \varepsilon^*$), the spatial profiles of the frequencies and rates of the processes that are determined by these parts of the EDF

Table 8.3. Ratios of the excitation and direct-ionization constants calculated with an EDF calculated with and without allowance for impacts of the second kind for $N_0 = 10^{17}$ cm^{-3}, $R_d = 1$ cm, $I = 10$ mA. From [11].

No.	Level	ε, eV	Ratio of the excitation constants
1	$3d_2^{7/2}$	14.01	1.5
2	$3d_2^{5/2}, 5s_2^{3/2}$	14.06	1.7
3	$3d_3^{5/2}, 5s_1^{3/2}$	14.09	1.9
4	$3d_1^{3/2}$	14.15	2.1
5	$5d_2^{'5/2}$	14.21	2.3
6	$3d_2^{'5/2}, 3d_2^{'5/2}, 5s_0^{'1/2}, 3s_1^{'5/2}$	14.23	2.4
7	$3d_1^{'3/2}$	14.30	2.7
8	$3d_1^{1/2}$	14.71	19
9	$6p_1^{'1/2}, 6p_1^{'3/2}, 6p_2^{'3/2}$	15.20	109
10	Ar$^+$	15.76	2670

can also be close to one another. This is illustrated by figure 8.13. It can be seen that the spatial profile of the direct-ionization frequency almost coincides with the profile of the step-ionization rate in spite of the large difference in the thresholds ($\epsilon_i = 15.76$ eV and $\epsilon_{st} = 4.35$ eV, respectively). We emphasize that this result is a consequence of the self-consistent simulation of a discharge. In this case, the frequency of direct ionization (Ar + e $\rightarrow$ Ar$^+$+e + e) is governed by EDF (8.14), which is formed by impacts of the second kind. For this reason, the $\nu_{idir}(r)$ profile is similar to the $n_m(r)$ $\nu_{st}(r)$ profile, although the step-ionization and direct-ionization processes themselves are unrelated.

The data presented above show that the local approximation is inapplicable to calculating the EDF even at relatively high pressures ($pR_d = 6$ cm Torr); i.e., it is almost always inapplicable under real conditions of a diffuse PC.

Experimental data are well coordinated with kinetic model PC. So for example, results of probe measurements of radial potential and EDF in low-pressure discharges have convincingly shown that the total energy really is an argument of EDF [15]. Measurements of a wall potential also show that results of the kinetic theory are in the good agreement with the experiment data. Thus, characteristics homogeneous in longitudinal direction PC are known reliably enough and can be calculated with the necessary accuracy.

8.3 The Langmuir paradox

To date, there still exists one of the most puzzling problems of the so-called Langmuir paradox, i.e., the reason why in low-pressure PCs (free-flight, $\lambda \gg R_d$) EDFs close to Maxwellian ones are observed. In the diffusion regime, when the electron mean free path $\lambda \ll R_d$, the collision processes, and the loss of electrons to the walls cause an approximately exponentially rapid fall-off of the EDF with different slopes in the corresponding energy intervals, which has been confirmed in numerous experiments and calculations. On the other hand, since the pioneering work of Langmuir in the 1920s [3], the physics of gas discharge plasmas was considered one of the most mystifying phenomena of the Maxwellian-like EDFs at low pressure (below 0.01–0.1 Torr cm). The linearity of probe current–voltage characteristics in semi-log plots indicated that the electron energy distribution function was close to Maxwellian. The radial gradient of the potential $\varphi(r)$, which exists in a plasma, traps electrons with energies below $e\Phi_w$. Contrary to expectations, however, as the energy increases at $\varepsilon > e\Phi_w$, the EDF does not fall off faster, even though the corresponding electrons can freely overcome the radial potential difference and, in a short free-flight time of the order $R_d/\sqrt{2\varepsilon/m}$, escape to the walls and recombine there. In 1925, Langmuir wrote, 'From the complete absence of a kink in the semilogarithmic plot at the wall potential we must conclude that the time of relaxation corresponding to the mechanism by which the electrons acquire their Maxwellian distribution is small compared to the time taken for the electrons to traverse the tube...' [3]. In the concluding 1930 review [16], he repeated, 'The mechanism by which these Maxwellian distributions are so quickly established in an ionized gas is not understood... in an ionized gas some additional and more effective agents must be chiefly responsible for the scattering of velocities.' Langmuir returned repeatedly to this paradoxical discrepancy between experiment and an estimate that seemed based on physically obvious considerations. In this regard, D Gabor wrote [17] of the 'worst discrepancy known to science.' The term 'Langmuir paradox,' itself, belongs to Gabor and has entered the scientific vocabulary.

A Maxwellian distribution usually develops in the course of random interactions (collisions) of particles among themselves or with a thermostat. Since inter-electronic collisions are rare under 'Langmuir paradox' conditions, various mechanisms for Maxwellization of the electrons have been discussed widely in the literature: references [17, 18, 19–29]. Right up to the present time, primary attention has been directed at searches for a universal mechanism for EDF Maxwellization, although these attempts have been, as yet, unsuccessful. Langmuir, himself, advanced the hypothesis that collective interactions of the electrons with turbulent electric fields that develop during plasma (Langmuir) oscillations may play a role. Gabor *et al* [17] observed some oscillations in the near-wall region and suggested that these oscillations are responsible for Maxwellization of the EDF. However, to date the presence of such oscillations (or at least their universal presence in low-pressure discharges) remains doubtful. Really, turbulent oscillations are not in thermodynamic equilibrium and cannot, in general, serve as a thermostat. Thus, interactions of the electrons with these oscillations, even if they occur, do not necessarily produce a Maxwellian electron distribution function. Other mechanisms for Maxwellization of the electron distribution function have been discussed, but no satisfactory explanation of the observed effects has yet been obtained. In addition, the influence of the oscillations on the EDF is caused by sufficiently complex processes, so it is absolutely not obvious that they precisely lead to the formation of the Maxwellian EDF. For example, electrons locked in the ambipolar potential well cannot reach the plasma periphery and interact with oscillations if the latter are present there. On the other hand, Langmuir's assumption that fast electrons moving away to the wall should lead to a kink in the EDF at $\varepsilon = e\Phi_w$ looks too rough in today's context. The point is that the characteristic time of their loss on the wall under condition $\lambda \gg R_d$ is not the transit time $t_f = R_d/\sqrt{2\varepsilon/m}$ but a much longer time in which the fast electron in elastic scattering gets to the small exit loss cone. This time, which exceeds even the time $1/\nu$ between collisions, tends to infinity at $\varepsilon = e\Phi_w$, smoothly decreasing with ε. Thus, the kink in the EDF at $e\Phi_w$, corresponding to the boundary between the locked and transit electrons, is smeared out and its experimental examination presents a difficult problem.

The Langmuir paradox itself was essentially proclaimed, since no analysis has been made of the EDF formation taking all the major factors into account. In [7] the hypothesis was formulated, that the Langmuir paradox is not related to some unknown mechanism for Maxwellization, but a combination of already known mechanisms may produce an electron energy distribution function that is close to an exponential with a constant slope (but to the physical features of the formation of the electron distribution function under these conditions). Specifically at low pressures, when electron mean energy is high, the account of the escape of the fast electrons to the tube walls would result to the exponential EDFs, close to the observed ones.

Because of the large difference between the momentum and energy relaxation times of the electrons, the electron distribution function now, as in the case $\lambda < R_d$, can be represented as the sum of isotropic $f_0(r, v)$ and of the small anisotropic $f_1(r, v, \theta)$ components. Since the anisotropy of the electron distribution function is caused by two independent factors, drift in the axial field E_z and radial escape of fast electrons to the vessel wall, the anisotropic EDF part $\overrightarrow{f_1}$ itself has two components. If the anisotropy is small, then they can be treated independently. For the first component, this is ensured by the smallness of the energy acquired along the mean free path in the field E_z compared to the total energy of the electron. Smallness of the second component, on the other hand, is associated with smallness of the loss cone within which the velocity of an electron leaving the plasma falls [30]. This condition is usually satisfied for most of the electrons of interest, whose energies moderately exceed the wall potential. Electrons with energies $\varepsilon < e\Phi_w$, the potential difference between the axis and wall of the vessel, are trapped in the volume by the ambipolar potential fall $e\Phi_a$ and the wall potential jump $e\Phi_s$, so that $\Phi_w = \Phi_a + \Phi_s$. Thus, the main part of the distribution function for the trapped electrons, f_0, depends only on the total energy $\varepsilon = w(r) + e\phi(r)$ (the kinetic $w = mv^2/2$ plus potential $e\phi(r)$ energies). Electrons with energies $\varepsilon > e\Phi_w$ can escape to the wall. It is equivalent to loss of electrons at a rate

$$\nu_w = 2\nu(\delta\Omega/4\pi). \tag{8.15}$$

Coefficient 2 shows up because of the symmetry of scattering at angles α and $\pi-\alpha$. As can be seen from equation (8.15), the electrons are lost much more slowly than through free-flight to the walls, which takes a time $R_d/\sqrt{2\varepsilon/m}$. In a cylindrical geometry, electrons with total energy $\varepsilon > e\Phi_w$ can escape to the wall if their perpendicular energy ε_p exceeds the sum of the potential and centrifugal energies at the wall, i.e. [7, 30, 31],

$$\varepsilon_p = mv_r^2/2 + \mu^2/2mr^2 + e\phi(r) \geqslant e\Phi_w + \mu^2/(2mR_d^2), \tag{8.16}$$

where

$$mv_r^2/2 = (\varepsilon - e\varphi(r))\cos^2\alpha.$$

As long as the loss cone is small ($\delta\Omega \ll 4\pi$), for isotropic elastic scattering of the electrons, their distribution function outside this cone, f_0, can also be regarded as isotropic and dependent only on ε. The character of the EDF anisotropy varies substantially, depending on whether it is associated with an axial field or with loss to the wall for isotropic scattering. In the first case, the EDF differs little from isotropic at all angles. Mathematically, this is reflected in the fact that the coefficients in the ordinary EDF expansion in terms of Legendre polynomials fall off rapidly with the harmonic number. In the second case, however, the EDF is almost isotropic outside a small anti-loss cone and practically zero inside it. In other words, the EDF has a discontinuity at its boundary. Outside this small cone, it is almost identical to the isotropic EDF part, f_0. Within the anti-loss cone, the distribution is also independent of angle. Therefore, all the coefficients in the expansion of the anisotropic EDF part in terms of spherical harmonics are also small (proportional $\delta\,\Omega$), but fall off with the harmonic number slowly. Thus, it is inappropriate to use this expansion and it is more convenient to consider $f_0(\varepsilon)$ and the electron distribution function inside the yield cone, $F(\varepsilon,r,\delta\Omega)$, separately.

After averaging over the volume, the kinetic equation for $f_0(\varepsilon)$ can be written in the form

$$\frac{\partial}{\partial\varepsilon}\bar{D}_{E}\frac{\partial f_0(\varepsilon)}{\partial\varepsilon} = \bar{\nu}_{ex}(\varepsilon)f_0(\varepsilon) - \bar{\nu}_{ex}(\varepsilon + \varepsilon_{ex})f_0(\varepsilon + \varepsilon_{ex})$$
$$+ \bar{\nu}_i(\varepsilon)f_0(\varepsilon) - 2\bar{\nu}_i(2\varepsilon + \varepsilon_i)f_0(2\varepsilon + \varepsilon_i) + \bar{\nu}_w f_0(\varepsilon) \tag{8.17}$$

where the coefficients of equation (8.17), obtained by averaging over the radius.

For simplicity, it is assumed that there is only one excitation process with a threshold ε_{ex} and only direct ionization from the ground state. The collision integral for ionization is written assuming that the kinetic energy is divided evenly between the incident and product electrons [31]. The anti-loss cone $\delta_1\Omega$ is essentially empty, since the corresponding electrons escape instantaneously to the wall (over a time $R_d/\sqrt{2\varepsilon/m}$) and, neglecting the effect of the radial electric field in the plasma on the fast electrons, the kinetic equation for the EDF $F(\varepsilon,r,\Omega)$ inside this cone can be written in the form

$$F(\varepsilon, x) = f_0(\varepsilon)\,(1 - \exp(-x/\lambda)) \tag{8.18}$$

Evidently, $F \ll f_0$.

In order to verify the accuracy of the approximate kinetic equation (8.17), it was compared in [7] with the results of an independent Monte Carlo method. For the conditions of interest to us, the most detailed work in this respect is [31], in which the EDF in the positive column of a glow discharge for an argon-like gas was modeled over a wide range of pressures 3 mTorr to 3 Torr. The authors were mainly guided by the collisional case ($\lambda > R_d$), for which they observed a change in the EDF slope for $\varepsilon > \varepsilon_{ex}$ owing to inelastic processes and the loss of electrons to the wall. A comparative analysis of these data [31] shows that when the pressure is lowered ($p < 10$ mTorr), the relative contribution of wall losses increases and there is a clear tendency for the EDF to flatten out to an exponential with a single slope throughout the entire energy range. In the following comparisons the cross-sections for argon, as in [31], are used.

The model calculations in reference [31] showed that the simple expression

$$\delta_1\Omega(\varepsilon, x) \approx 2\pi\left(1 - \sqrt{\frac{\Phi_w - \phi(x)}{\varepsilon/e - \phi(x)}}\right) \tag{8.19}$$

in the case of a plane geometry is a good approximation for a cylinder, as well. As a comparison, figure 8.14 shows f_0 calculated according to equation (8.17) for the

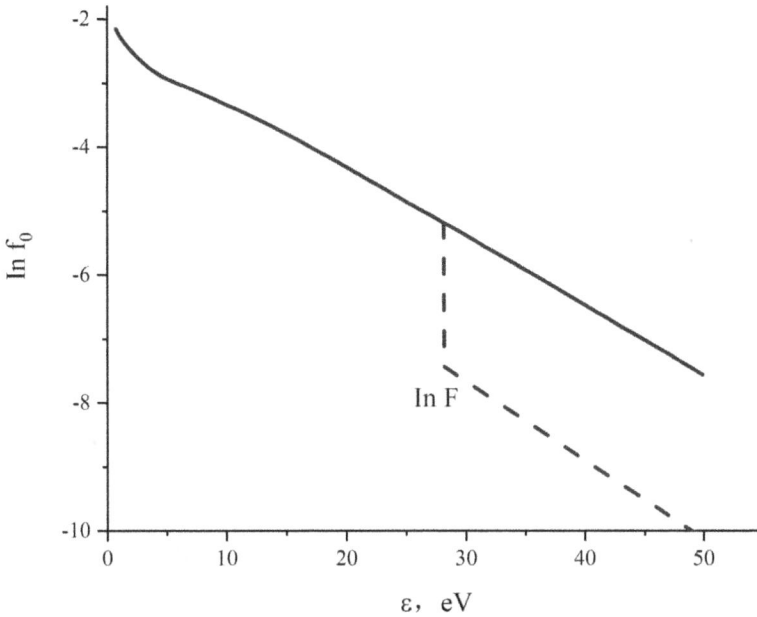

Figure 8.14. $f_0(\varepsilon)$ calculated according to equation (8.16). $N = 10^{14}$ cm^{-3}, the vessel radius $R_d = 1$ cm, $\phi_w = 28.5$ V, and $E_z = 1$ V cm^{-1}. From [7].

conditions of [31] (neutral gas density $N = 10^{14}$ cm^{-3} and cylinder radius $R_d = 1$ cm). The self-consistent values of the wall potential $\Phi_w = 28.5$ V and longitudinal field $E_z = 1$ V cm^{-1} turned out to be close to those calculated in [31] (30 eV and 1.02 V cm^{-1}, respectively). A semi-log plot of the EDF is clearly close to linear with an effective electron temperature $T_e = 9.3$ eV. Significant deviations from a Maxwellian EDF are observed only at low energies. These calculations also confirm the well known fact, that T_e in low-pressure discharges is of the order of the threshold ε_{ex} for inelastic processes. Physically, such high temperatures are necessary in order for an electron to complete an ionization event during the short ion mean free time τ_i to reach the wall. If we use the EDF, figure 8.14, to construct a plot of the probe current

$$I(eV) = \int_{eV}^{\infty} (\varepsilon - eV) f_0(\varepsilon) d\varepsilon \qquad (8.20)$$

from which the Langmuir paradox was first observed in the literature, then we obtain an almost straight line which is essentially identical to an exponential (figure 8.15). This is because an integral dependence of the type of equation (8.20) has a weaker dependence on the details of the EDF behavior. It is known that the finite magnitude of the differentiated signal at the probe means that the measured EDF is significantly distorted at energies near the plasma potential for small potentials, which are around $(0.3 - 1)T_e$ in the experiment [7]. The calculations [7] also show that curves analogous to those of figures 8.14 and 8.15 are obtained for pure gases when $\lambda > R_d$ under all these conditions.

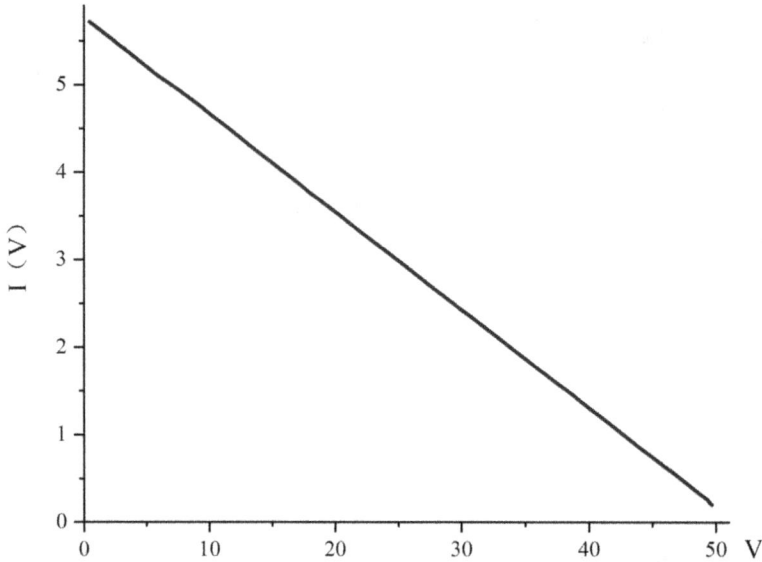

Figure 8.15. The probe current I as a function of the potential V. The conditions are the same as in figure 8.14. From [7].

In order to understand the reasons for such surprising behavior of f_0, in [7] the analyses of the terms in the kinetic equation (8.17) for the considered conditions was performed. Obviously, for $\varepsilon > e\Phi_w$ the rate of loss to the wall significantly exceeds the inelastic collision rates, while the energy dependences of $\bar{D}_E(\varepsilon)$ and $\bar{\nu}_w(\varepsilon)$ are similar. Here the kinetic equation (8.17) for the fast part of the electron energy distribution function with $\varepsilon > e\Phi_w$ reduces to

$$\frac{\partial}{\partial \varepsilon} \bar{D}_E \frac{\partial f_0(\varepsilon)}{\partial \varepsilon} = \bar{\nu}_w f_0(\varepsilon) \tag{8.21}$$

and, roughly speaking, the electron 'temperature' for the fast part of the electron energy distribution function,

$$T_w = \sqrt{\bar{D}_E/\bar{\nu}_w}, \tag{8.22}$$

is practically constant and independent of the energy. The reason for this is the weak, decreasing energy dependence of the elastic cross-section at high energies. This leads to a small energy dependence for $\nu(w)$ and, therefore, for T_w in equation (8.21). It is well known that at high energies $\varepsilon \geqslant \varepsilon_i$ the elastic scattering cross-sections fall off slowly with increasing energy and behave this way for almost all elements. Thus, this sort of dependence is quite universal. At high energies, therefore, T_w (equation (8.21)), is nearly constant for essentially any pure gas and the EDF for $\varepsilon > e\Phi_w$ should be close to Maxwellian,

$$f_0(\varepsilon) \approx \exp(-\varepsilon/T_w). \tag{8.23}$$

We now consider some possible reasons why the EDF may have the same slope in the elastic energy range, $\varepsilon < \varepsilon_{ex}$, as well. At first glance, this seems paradoxical, since it is well known that the actual energy dependences of $\nu(w)$ are different in this region for different gases. Let us write down a model solution of the kinetic equation (8.17) for $\varepsilon < \varepsilon_{ex}$ with the fast part of the EDF in the form (8.23)

$$f_{0m}(\varepsilon) = \Phi_1 - \Phi_2 = (\exp(-\varepsilon_{ex}/T_w)$$
$$+ \exp(-2\varepsilon_{ex}/T_w))(1 + (\varepsilon_{ex} - \varepsilon)/T_w) - \exp(-(\varepsilon + \varepsilon_{ex})/T_w) \tag{8.24}$$

which represents the difference between the linear Φ_1 and exponential Φ_2 functions. In the elastic energy range, the model EDF (8.24) is evidently never Maxwellian, even for a Maxwellian fast part (8.23). The functions (8.23) and (8.24) are close to one another for $\varepsilon \sim \varepsilon_{ex}$ and diverge with decreasing energy. Since their differences are greatest at low energies, for $\varepsilon/T_e \ll 1$, on expanding the exponent in Φ_2 in a Taylor series, near zero we obtain

$$f_0(\varepsilon \approx 0) \approx (\varepsilon_{ex}/T_w)\exp(-\varepsilon_{ex}/T_w). \tag{8.25}$$

The expression on the right of equation (8.25) is close to unity (i.e., to the corresponding values of equation (8.23) only for $\varepsilon_{ex} \approx T_w$ and falls off sharply with decreasing T_w. Thus, the closeness of the approximate EDF in the elastic energy range to an exponential with the temperature of the fast portion of the electron distribution is determined by the parameter ε_{ex}/T_w. This sort of approximation is possible only for high $T_w \approx \varepsilon_{ex}$. This is illustrated in figure 8.16, which shows the model EDFs for $\varepsilon_{ex}/T_e = 0.5, 1$, and 5. Evidently, the complete EDF should never be

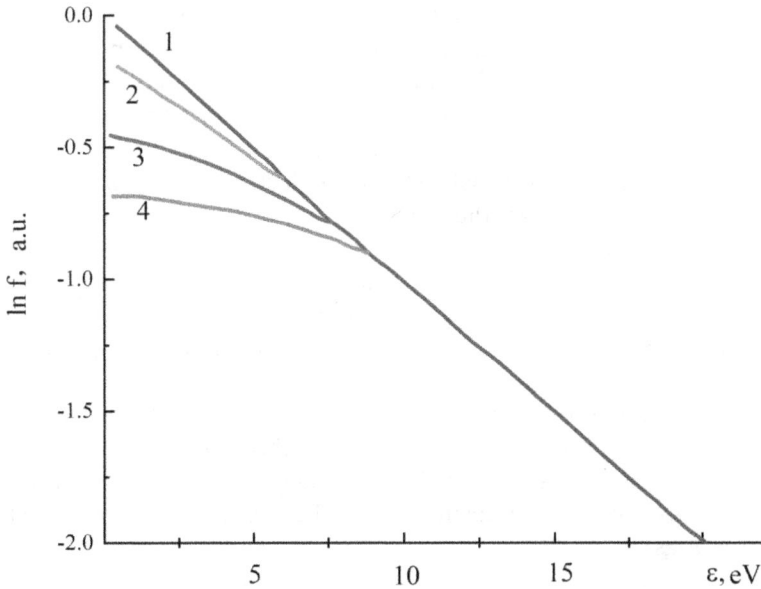

Figure 8.16. Model electron energy distribution functions (8.24) for $\varepsilon_{ex}/T_e = 5$ (2), 1 (3), and 0.5 (4) and the electron energy distribution function of equation (8.23) (1). From [7].

strictly Maxwellian. In [7] it was shown that it can be approximated by an exponential with a single slope only when the electron temperature is high and close to the inelastic threshold. Therefore, in order for the entire EDF in low-pressure discharges to be approximated by a Maxwellian distribution, two major conditions must be satisfied: the elastic cross-section must fall off slowly at high energies (beyond the ionization potential of the working gas) and the electron temperature must be high (of the order of the inelastic threshold). Failure of either of these conditions will cause the EDF to deviate significantly from an exponential with a single slope, i.e., the Langmuir paradox would be absent.

Recent work [32] has provided some EDF measurements under Langmuir paradox conditions. The non-Maxwellian EEDFs in argon gas found in those experiments cast doubts on the existence of the Langmuir paradox (in any case, on its universality), and call for further experiments. Thus, the Langmuir paradox remains a topical and fundamental problem in the physics of gas discharges, which must be solved in order to understand the properties of low-temperature plasmas and to justify the validity of the probe diagnostics used on them.

8.4 PC in electronegative gases

In the presence of negative ions, the processes of spatial transport, which determine the density profiles and other plasma parameters, possess a number of specific features (see [14] for details). Early attempts to reduce the problem to a set of ambipolar diffusion coefficients by using simplified models were contradictory and there were no criteria for their applicability. In [33–36], it was shown that a specific feature of an electronegative-gas plasma with $T_e \gg T_i$ is that it stratifies into regions with different ion compositions separated by a sharp boundary. In the external region (shell) of such a plasma, negative ions are practically absent (figures 8.17–8.19), because they are drawn by the ambipolar electric field into the plasma interior. The presence of this shell is of fundamental importance because it confines the negative ions inside the plasma volume. As a result, the flux of negative ions to the wall is practically absent (in contrast to those of electrons and positive ions). In such a situation, the only means to extract negative ions from the discharge is to apply an accelerating voltage U to the wall (or an extracting electrode). The magnitude of this voltage should be large enough for the space charge layer produced at the plasma boundary to extend to the inner region containing negative ions. The thicker the shell, the higher the voltage ($U \sim L_{sh}^{2/3}$) that must be applied to enable the flux of negative ions to the wall.

For the sake of qualitative analysis, we would consider a plasma consisting of only electrons, positive ions, and negative ions (subscripts e, p and n, respectively). To explain the dependences observed and predict how they are affected by the external conditions, we consider, as in [33–36], the conventional set of the fluid drift–diffusion equations

$$-D_p \nabla (\nabla n_p + k n_p \nabla n_e / n_e) = \nu_i n_e - K_r n_n n_p, \tag{8.26}$$

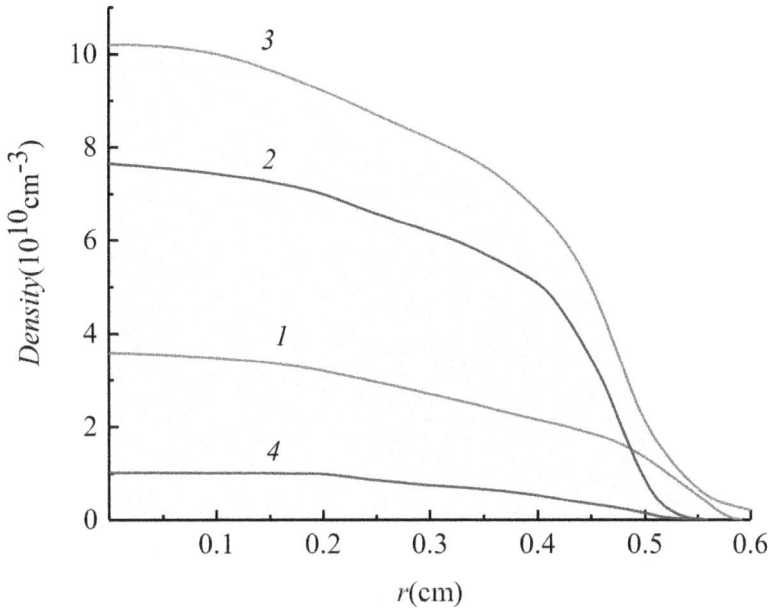

Figure 8.17. Profiles of the charged particle densities for $p = 1$ Torr and $I = 50$ mA: (1) n_e, (2) n_n, (3) n_p, and (4) $n[O^+]$. From [10].

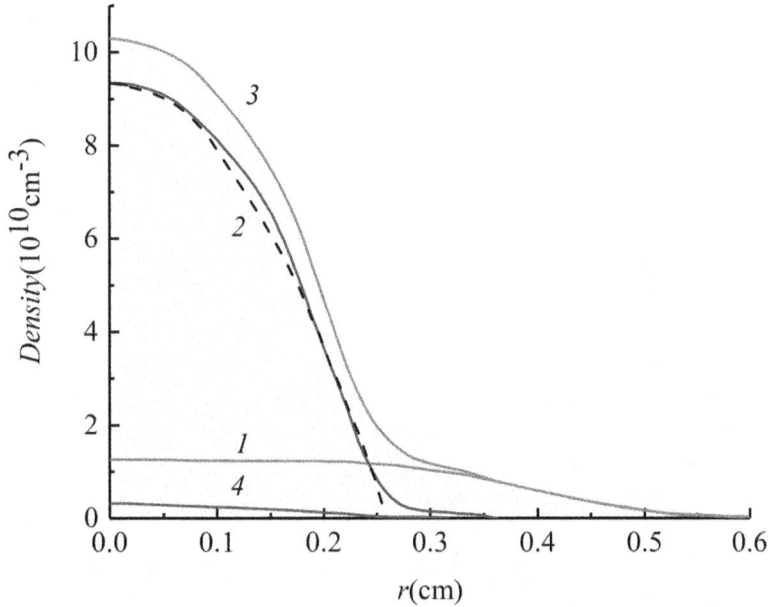

Figure 8.18. The same as in figure 8.17 for $p = 0.15$ Torr without allowance for ion heating: (1) n_e, (2) n_n, (3) n_p, and (4) $n[O+]$. The dashed curve shows parabolic distribution (8.49). From [10].

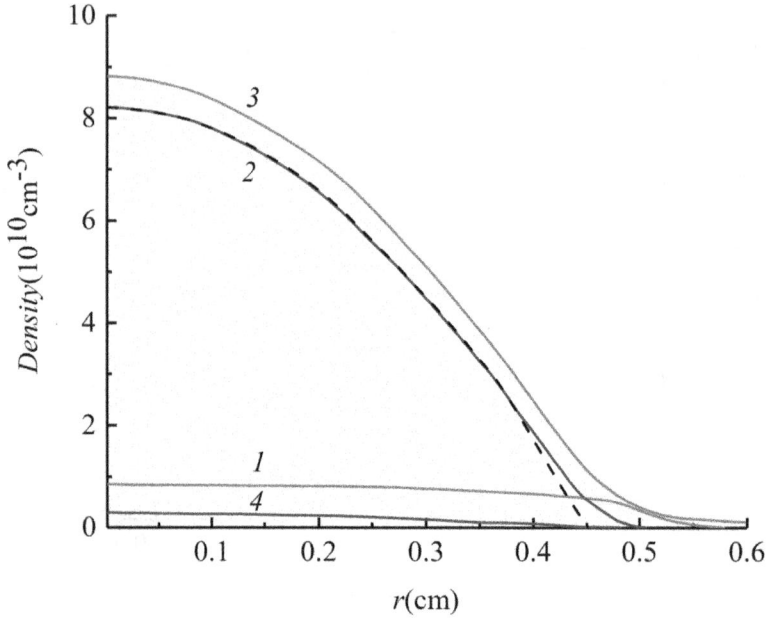

Figure 8.19. The same as in figure 8.18, but with allowance for ion heating. From [10].

$$-D_n \nabla (\nabla n_n - k n_n \nabla n_e / n_e) = \nu_a n_e - \nu_d n_n - K_r n_n n_p, \tag{8.27}$$

$$n_p = n_n + n_e \tag{8.28}$$

with a Boltzmann distribution for the electrons: $E = -T_e \nabla n_e / n_e$. Here, ν_i, ν_a, and ν_d are the ionization, attachment, and detachment frequencies, respectively; K_r is the rate constant for ion–ion recombination; and $k = T_e / T_i$ is the electron-to-ion temperature ratio.

The boundary conditions for the set of equations (8.26) and (8.27) are [14, 37]

$$\nabla n_n = \nabla n_p = 0 \text{ at } r = 0, \quad n_n = n_p = \nabla n_n = 0 \text{ at } r = R_d. \tag{8.29}$$

We illustrate these considerations on full-scale kinetic simulations of PC plasma in [10, 38] in which we compared the results of kinetic and fluid simulations of the positive column plasma of a dc oxygen discharge in a 12 mm diameter glass tube at pressures of 0.05–3 Torr and discharge currents of 5–200 mA. The density and mean energy of the electron component can be obtained by solving either fluid balance equations or the kinetic equation for the electron distribution function (EDF). The self-consistent electric field is found from Poisson's equation. Heavy particles are described in the fluid model.

The accounted volume plasma-chemical processes with the participation of various atomic and molecular oxygen states are listed in table 8.4. In the fluid model, the rate constants of the processes with the participation of electrons were obtained by convoluting the corresponding cross-sections with a Maxwellian EDF,

Table 8.4. Volume plasma-chemical processes involved in simulations. From [38].

No.	Reaction	$\Delta\varepsilon$, eV	Rate constant
			Elastic electron scattering
1	$e + O_2 \longrightarrow e + O_2$	0	Cross-section
2	$e + O_2(a^1\Delta) \longrightarrow e + O_2(a^1\Delta)$	0	Cross-section
3	$e + O_2(b^1\Sigma) \longrightarrow e + O_2(b^1\Sigma)$	0	Cross-section
4	$e + O_2(v_1) \longrightarrow e + O_2(v_1)$	0	Cross-section
5	$e + O_2(Ry) \longrightarrow e + O_2(Ry)$	0	Cross-section
6	$e + O \longrightarrow e + O$	0	Cross-section
7	$e + O(^1D) \longrightarrow e + O(^1D)$	0	Cross-section
8	$e + O(^1S) \longrightarrow e + O(^1S)$	0	Cross-section
9	$e + O_3 \longrightarrow e + O_3$	0	Cross-section
			Inelastic processes with the participation of electrons
10	$e + O_2 \longrightarrow O_- + O$	3.637	Cross-section
11	$e + O_2 \longrightarrow e + O_2(v_1)$	0.19	Cross-section
12	$e + O_2(v_1) \longrightarrow e + O_2$	−0.19	Cross-section
13	$e + O_2 \longrightarrow e + O_2(v_2)$	0.38	Cross-section
14	$e + O_2 \longrightarrow e + O_2(v_3)$	0.57	Cross-section
15	$e + O_2 \longrightarrow e + O_2(v_4)$	0.75	Cross-section
16	$e + O_2 \longrightarrow e + O_2(a^1\Delta)$	0.97	Cross-section
17	$e + O_2(a^1\Delta) \longrightarrow e + O_2$	−0.97	Obtained from a detailed balance with c 16
18	$e + O_2 \longrightarrow e + O_2(b^1\Sigma)$	1.63	Cross-section
19	$e + O_2(b^1\Sigma) \longrightarrow e + O_2$	−1.63	Obtained from a detailed balance with c 18
20	$e + O_2 \longrightarrow e + 2O$	5.12	Cross-section
21	$e + O_2 \longrightarrow e + O + O(^1D)$	7.1	Cross-section
22	$e + O_2 \longrightarrow 2e + O_{2+}$	12.6	Cross-section
23	$e + O_2 \longrightarrow 2e + O + O_+$	18.8	Cross-section
24	$e + 2O_2 \longrightarrow O_2 + O_{2-}$	−5.03	$k_{24} = 3.6E - 43 T_e^{-0.5} \mathrm{m^6 s^{-1}}$
25	$e + O_{2+} \longrightarrow 2O$	−6.96	Cross-section
26	$e + O_{2+} \longrightarrow O + O(^1D)$	−5.0	Cross-section
27	$e + O_2(a^1\Delta) \longrightarrow 2e + O_{2+}$	−11.63	Cross-section
28	$e + O_2(b^1\Sigma) \longrightarrow 2e + O_{2+}$	−10.97	$k_{28} = 1.3E - 15 T_e^{-1.1} \exp(-10.43/T_e) \ \mathrm{m^3 s^{-1}}$
29	$e + O_3 \longrightarrow O_- + O_2$	−0.42	Cross-section
30	$e + O_3 \longrightarrow O + O_{2-}$	0.60	Cross-section
31	$e + O \longrightarrow e + O(^1D)$	1.97	Cross-section
32	$e + O \longrightarrow e + O(^1S)$	4.24	Cross-section
33	$e + O(^1S) \longrightarrow e + O$	−4.24	Obtained from a detailed balance with c 32
34	$e + O \longrightarrow 2e + O_+$	13.67	Cross-section
35	$e + O(^1D) \longrightarrow e + O$	−1.97	Cross-section
36	$e + O(^1D) \longrightarrow 2e + O_+$	11.7	Cross-section
37	$e + O(^1S) \longrightarrow 2e + O_+$	9.43	$k_{37} = 6.6E - 15 T_e^{0.6} \exp(-9.43/T_e) \ \mathrm{m^3 s^{-1}}$
38	$e + O_- \longrightarrow 2e + O$	1.53	$k_{38} = 1.95E - 18 T_e^{0.5} \exp(-3.4/T_e) \ \mathrm{m^3 s^{-1}}$

39	$e + O_+ \longrightarrow O(^1D)$	-11.7	$k_{39} = 5.3E - 19T_e^{-0.5}\text{m}^3\text{s}^{-1}$
40	$2e + O_+ \longrightarrow 2e + O(^1D)$	-11.7	$k_{40} = 5.12E - 36T_e^{-4.5}\text{m}^6\text{s}^{-1}$
41	$e + O \longrightarrow e + O(3s^5S_0)$	9.15	Cross-section
42	$e + O \longrightarrow e + O(3s^3S_0)$	9.51	Cross-section
43	$e + O \longrightarrow e + O(3p^5P_0)$	10.73	Cross-section
44	$e + O \longrightarrow e + O(3p^3P_0)$	10.98	Cross-section
45	$e + O_2 \longrightarrow e + O_2(Rot)$	0.02	Cross-section
46	$e + O_2 \longrightarrow e + O_2(v_5)$	0.19	Cross-section
47	$e + O_2 \longrightarrow e + O_2(v_6)$	0.38	Cross-section
48	$e + O_2 \longrightarrow e + O_2(^1\Pi_g)$	8.4	Cross-section
49	$e + O_2 \longrightarrow e + O_2(^1a\Sigma_u^+)$	10.0	Cross-section
50	$e + O_2 \longrightarrow e + O_2 + h\nu$ (130 nm)	9.547	Cross-section
51	$e + O_2(a^1\Delta) \longrightarrow e + O_2(b^1\Sigma)$	0.65	Cross-section
52	$e + O_2(b^1\Sigma) \longrightarrow e + O_2(a^1\Delta)$	-0.65	Cross-section
53	$e + O_2 \longrightarrow e + O_2(Ry)$	4.47	Cross-section
54	$e + O_2(Ry) \longrightarrow e + O_2$	-4.47	Cross-section
55	$e + O_2(a^1\Delta) \longrightarrow e + O_2(Ry)$	3.45	Cross-section
56	$e + O_2 \longrightarrow e + O + O(^1S)$	9.36	Cross-section
57	$e + O_2(a^1\Delta) \longrightarrow O + O_-$	2.57	Cross-section
58	$e + O_{2+} \longrightarrow O_2(Ry)$	-7.66	Cross-section
59	$e + O_2 + O_3 \longrightarrow O_2 + O_{3-}$	-0.679	$4.6E - 40 \text{ m}^3\text{s}^{-1}$
60	$e + O_{2+} \longrightarrow O + O(^1S)$	-2.73	$2.42E - 13T_e^{-0.55}\text{m}^3\text{s}^{-1}$
61	$e + O_{4+} \longrightarrow 2O_2$	-0.8	$2.42E - 11T_e^{-0.5}\text{m}^3\text{s}^{-1}$
62	$e + O_{4+} \longrightarrow O_2 + O_2(Ry)$	3.68	$2.425E - 12T_e^{-0.5}\text{m}^3\text{s}^{-1}$

Reactions between heavy species

63	$e + O_{2+} \longrightarrow O + O_2$	$k_{63} = 5.69E - 11T_g^{-1}\text{m}^3\text{s}^{-1}$
64	$O_- + O_{2+} \longrightarrow 3O$	$k_{64} = 1E - 13 \text{ m}^3\text{s}^{-1}$
65	$O_- + O_+ \longrightarrow 2O$	$k_{65} = 5.96E - 11T_g^{-1}\text{m}^3\text{s}^{-1}$
66	$O_{2-} + O_{2+} \longrightarrow 2O_2$	$k_{66} = 5.96E - 11T_g^{-1}\text{m}^3\text{s}^{-1}$
67	$O_{2-} + O_{2+} \longrightarrow O_2 + 2O$	$k_{67} = 1E - 13 \text{ m}^3\text{s}^{-1}$
68	$O_+ + O_{2-} \longrightarrow O_2 + O$	$k_{68} = 5.96E - 11T_g^{-1}\text{m}^3\text{s}^{-1}$
69	$O_{2+} + O_{3-} \longrightarrow O_2 + O_3$	$k_{69} = 5.96E - 11T_g^{-1}\text{m}^3\text{s}^{-1}$
70	$O_{2+} + O_{3-} \longrightarrow 2O + O_3$	$k_{70} = 1E - 13 \text{ m}^3 \text{ s}^{-1}$
71	$O_+ + O_{3-} \longrightarrow O + O_3$	$k_{71} = 5.96E - 11T_g^{-1}\text{m}^3\text{s}^{-1}$
72	$O_- + O_{2+} + O_2 \longrightarrow O + 2O_2$	$k_{72} = 3.066E - 31T_g^{-2.5}\text{m}^3\text{s}^{-1}$
73	$O_- + O_+ + O_2 \longrightarrow 2O + O_2$	$k_{73} = 3.066E - 31T_g^{-2.5}\text{m}^3\text{s}^{-1}$
74	$O + O_- \longrightarrow O_2 + e$	$k_{74} = 1.159E - 17T_g^{0.5} \text{ m}^3\text{s}^{-1}$
75	$O_- + O_2(a^1\Delta) \longrightarrow O_3 + e$	$k_{75} = 1.738E - 17T_g^{0.5} \text{ m}^3\text{s}^{-1}$
76	$O_- + O_2(b^1\Sigma) \longrightarrow O_2 + O + e$	$k_{76} = 4E - 17T_g^{0.5} \text{ m}^3\text{s}^{-1}$
77	$O_- + O_2 \longrightarrow O_3 + e$	$k_{77} = 2.896E - 22T_g^{0.5} \text{ m}^3\text{s}^{-1}$
78		

(Continued)

Table 8.4. (*Continued*)

No.	Reaction	$\Delta\varepsilon$, eV	Rate constant
	$O_- + O_3 \longrightarrow 2O_2 + e$		$k_{78} = 1.744E - 17T_g^{0.5}$ m^{3}s^{-1}
79	$O_- + O_3 \longrightarrow O + O_{3-}$		$k_{79} = 1.153E - 17T_g^{0.5}$ m^{3}s^{-1}
80	$O_- + O_3 \longrightarrow O_2 + O_{2-}$		$k_{80} = 5.509E - 19T_g^{0.5}$ m^{3}s^{-1}
81	$O + O_{2-} \longrightarrow O + O_{2-}$		$k_{81} = 8.69E - 18T_g^{0.5}$ m^{3}s^{-1}
82	$O + O_{2-} \longrightarrow O_3 + e$		$k_{82} = 8.69E - 18T_g^{0.5}$ m^{3}s^{-1}
83	$O_2(a^1\Delta) + O_{2-} \longrightarrow 2O_2 + e$		$k_{83} = 1.159E - 17T_g^{0.5}$ m^{3}s^{-1}
84	$O_{2-} + O_3 \longrightarrow O_{2-} + O_3$		$k_{84} = 3.746E - 17T_g^{0.5}$ m^{3}s^{-1}
85	$O + O_{3-} \longrightarrow O_2 + O_{2-}$		$k_{85} = 1.448E - 17T_g^{0.5}$ m^{3}s^{-1}
86	$O + O_+ + O_2 \longrightarrow O_2 + O_{2+}$		$k_{86} = 5.793E - 43T_g^{0.5}$ m^{3}s^{-1}
87	$O_+ + O_2 \longrightarrow O + O_{2+}$		$k_{87} = 1.953E - 16T_g^{0.4}$ m^{3}s^{-1}
88	$O_+ + O_3 \longrightarrow O_2 + O_{2+}$		$k_{88} = 1E - 16$ m^{3}s^{-1}
89	$O(^1D) + O_2 \longrightarrow 2O$		$k_{89} = 1E - 18$ m^{3}s^{-1}
90	$O(^1D) + O_2 \longrightarrow O + O_2(b^1\Sigma)$		$k_{90} = 2.56E - 17\exp(+ 67/T_g)$ m^3 s^{-1}
91	$O(^1D) + O_2 \longrightarrow O + O_2(a^1\Delta)$		$k_{91} = 1.6E - 18\exp(+ 67/T_g)$ m^3 s^{-1}
92	$O(^1D) + O_2 \longrightarrow O + O_2$		$k_{92} = 4.8E - 18\exp(+ 67/T_g)$ m^3 s^{-1}
93	$O(^1D) + O_3 \longrightarrow 2O + O_2$		$k_{93} = 1.2E - 16$ m^3 s^{-1}
94	$O(^1D) + O_3 \longrightarrow 2O_2$		$k_{94} = 1.2E - 16$ m^3 s^{-1}
95	$O(^1S) + O_2 \longrightarrow O(^1D) + O_2$		$k_{95} = 3.2E - 16\exp(- 850/T_g)$ m^3 s^{-1}
96	$O(^1S) + O_2 \longrightarrow O + O_2$		$k_{96} = 1.6E - 18\exp(- 850/T_g)$ m^3 s^{-1}
97	$O(^1S) + O_2(a^1\Delta) \longrightarrow O + O_2$		$k_{97} = 1.1E - 16$ m^3 s^{-1}
98	$O(^1S) + O_2(a^1\Delta) \longrightarrow O(^1D) + O_2(b^1\Sigma)$		$k_{98} = 2.9E - 17$ m^3 s^{-1}
99	$O(^1S) + O_2(a^1\Delta) \longrightarrow 3O$		$k_{99} = 3.2E - 17$ m^3 s^{-1}
100	$O(^1S) + O \longrightarrow O(^1D) + O$		$k_{100} = 1.67E - 17\exp(- 300/T_g)$ m^3 s^{-1}
101	$O(^1S) + O \longrightarrow 2O$		$k_{101} = 3.33E - 17\exp(- 300/T_g)$ m^3 s^{-1}
102	$O(^1S) + O_3 \longrightarrow 2O$		$k_{102} = 5.8E - 16$ m^3 s^{-1}
103	$O_2(a^1\Delta) + O \longrightarrow O_2 + O$		$k_{103} = 2E - 22$ m^3 s^{-1}
104	$O_2(a^1\Delta) + O_2 \longrightarrow 2O_2$		$k_{104} = 3E - 24\exp(- 200/T_g)$ m^3 s^{-1}
105	$2O_2(a^1\Delta) \longrightarrow O_2$		$k_{105} = 9E - 23\exp(- 560/T_g)$ m^3 s^{-1}
106	$2O_2(a^1\Delta) \longrightarrow O_2 + O_2(b^1\Sigma)$		$k_{106} = 9E - 23\exp(- 560/T_g)$ m^3 s^{-1}
107	$2O_2(a^1\Delta) + O_2 \longrightarrow 2O_3$		$k_{107} = 1E - 43\exp(- 560/T_g)$ m^3 s^{-1}
108	$2O_2(a^1\Delta) + O_2 \longrightarrow 2O_3$		$k_{108} = 1.709E - 28T_g$ m^3 s^{-1}
109	$O_2(a^1\Delta) + O_3 \longrightarrow 2O_2 + O$		$k_{109} = 5.2E - 17\exp(- 2840/T_g)$ m^3 s^{-1}
110	$2O_2(b^1\Sigma) \longrightarrow O_2(a^1\Delta) + O_2$		$k_{110} = 2.085E - 24T_g^{0.5}$ m^{3}s^{-1}
111	$O_2(b^1\Sigma) + O_2 \longrightarrow O_2(a^1\Delta) + O_2$		$k_{111} = 2.085E - 25T_g^{0.5}$ m^{3}s^{-1}
112	$O_2(b^1\Sigma) + O_2 \longrightarrow 2O_2$		$k_{112} = 2.317E - 28T_g^{0.5}$ m^{3}s^{-1}
113	$O_2(b^1\Sigma) + O \longrightarrow O_2(a^1\Delta) + O$		$k_{113} = 4.171E - 21T_g^{0.5}$ m^{3}s^{-1}
114	$O_2(b^1\Sigma) + O \longrightarrow O_2 + O$		$k_{114} = 4.634E - 22T_g^{0.5}$ m^{3}s^{-1}
115	$O_2(b^1\Sigma) + O_3 \longrightarrow 2O_2 + O$		$k_{115} = 4.246E - 19T_g^{0.5}$ m^{3}s^{-1}
116	$O_2(b^1\Sigma) + O_3 \longrightarrow O_2(a^1\Delta) + O_3$		$k_{116} = 4.246E - 19T_g^{0.5}$ m^{3}s^{-1}
117	$O_2(b^1\Sigma) + O_3 \longrightarrow O_2 + O_3$		$k_{117} = 4.246E - 19T_g^{0.5}$ m^{3}s^{-1}

118 $O_2(v_1) + O \longrightarrow O_2 + O$ $k_{118} = 5.793E - 22T_g^{0.5} \text{ m}^3\text{s}^{-1}$

119 $O_2(v_1) + O_2 \longrightarrow 2O_2$ $k_{119} = 5.793E - 22T_g^{0.5} \text{ m}^3\text{s}^{-1}$

120 $2O + O_2 \longrightarrow 2O_2$ $k_{120} = 9.268E - 45T_g^{-0.63}\text{m}^6\text{s}^{-1}$

121 $3O \longrightarrow O + O_2$ $k_{121} = 3.334E - 44T_g^{-0.63}\text{m}^6\text{s}^{-1}$

122 $2O + O_2 \longrightarrow O_2(a^1\Delta) + O_2$ $k_{122} = 6.987E - 46T_g^{-0.63}\text{m}^6\text{s}^{-1}$

123 $3O \longrightarrow O_2(a^1\Delta) + O$ $k_{123} = 2.509E - 45T_g^{-0.63}\text{m}^6\text{s}^{-1}$

124 $O + 2O_2 \longrightarrow O_3 + O_2$ $k_{124} = 5.081E - 39T_g^{-0.28}\text{m}^6\text{s}^{-1}$

125 $2O + O_2 \longrightarrow O + O_3$ $k_{125} = 3.166E - 43T_g^{-1.2}\text{m}^6\text{s}^{-1}$

126 $O + O_3 \longrightarrow 2O_2$ $k_{126} = 8E - 18\exp(-2060/T_g) \text{ m}^3 \text{ s}^{-1}$

127 $O_2 + O_3 \longrightarrow O + 2O_2$ $k_{127} = 1.56E - 15\exp(-11\,900/T_g) \text{ m}^3 \text{ s}^{-1}$

128 $O_2(Ry) \longrightarrow O_2$ $k_{128} = 0.015 \text{ s}^{-1}$

129 $O_2 + O_2(Ry) \longrightarrow O_2 + O_2(a^1\Delta)$ $k_{129} = 1.86E - 19 \text{ m}^3 \text{ s}^{-1}$

130 $O_2 + O_2(Ry) \longrightarrow O_2 + O_2(b^1\Sigma)$ $k_{130} = 1.86E - 19 \text{ m}^3 \text{ s}^{-1}$

131 $O(^1D) + O_2(a^1\Delta) \longrightarrow O_2 + O$ $k_{131} = 1E - 17 \text{ m}^3 \text{ s}^{-1}$

132 $O(^1S) + O_2 \longrightarrow O + O_2(a^1\Delta)$ $k_{132} = 1.5E - 18\exp(-850/T_g) \text{ m}^3 \text{ s}^{-1}$

133 $O(^1S) + O_2 \longrightarrow O + O_2(b^1\Sigma)$ $k_{133} = 7.3E - 19\exp(-850/T_g) \text{ m}^3 \text{ s}^{-1}$

134 $O(^1S) + O_2 \longrightarrow O + O_2(Ry)$ $k_{134} = 7.3E - 19\exp(-850/T_g) \text{ m}^3 \text{ s}^{-1}$

135 $O(^1S) + O_2(a^1\Delta) \longrightarrow O + O_2(Ry)$ $k_{135} = 1.3E - 16 \text{ m}^3 \text{ s}^{-1}$

136 $O_2(b^1\Sigma) + O_3 \longrightarrow O_2(a^1\Delta) + O_3$ $k_{136} = 7.1E - 17 \text{ m}^3 \text{ s}^{-1}$

137 $O + O_2 + O_2(a^1\Delta) \longrightarrow O_2(b^1\Sigma) + O_3$ $k_{137} = 1.56E - 40T_g^{-1.5}\text{m}^6\text{s}^{-1}$

138 $O + O_2 + O_2(a^1\Delta) \longrightarrow O + 2O_2$ $k_{138} = 3E - 44 \text{ m}^6 \text{ s}^{-1}$

139 $O + O_3 \longrightarrow O_2 + O_2(a^1\Delta)$ $k_{139} = 2.4E - 19 \exp(-2060/T_g) \text{ m}^3\text{s}^{-1}$

140 $O + O_3 \longrightarrow O_2 + O_2(b^1\Sigma)$ $k_{140} = 8E - 20\exp(-2060/T_g) \text{ m}^3 \text{ s}^{-1}$

141 $2O_3 \longrightarrow O + O_2 + O_3$ $k_{141} = 1.65E - 15\exp(-11\,435/T_g) \text{ m}^3 \text{ s}^{-1}$

142 $2O + O_2 \longrightarrow O_2 + O_2(Ry)$ $k_{142} = 1.2E - 46 \text{ m}^6 \text{ s}^{-1}$

143 $2O + O_2 \longrightarrow O_2 + O_2(b^1\Sigma)$ $k_{143} = 7.6E - 44T_g^{-1}\exp(-170/T_g) \text{ m}^6 \text{ s}^{-1}$

144 $O + O_2 + O_3 \longrightarrow 2O_3$ $k_{144} = 1.3E - 41T_g^{-2}\text{m}^6\text{s}^{-1}$

145 $2O_2 + O_{2+} \longrightarrow O_2 + O_{4+}$ $k_{145} = 1.25E - 38T_g^{-1.5}\text{m}^6\text{s}^{-1}$

146 $O_2(a^1\Delta) + O_{4+} \longrightarrow 2O_2 + O_{2+}$ $k_{146} = 1E - 16 \text{ m}^3 \text{ s}^{-1}$

147 $O_2(b^1\Sigma) + O_{4+} \longrightarrow 2O_2 + O_{2+}$ $k_{147} = 1E - 16 \text{ m}^3 \text{ s}^{-1}$

148 $O + O_{4+} \longrightarrow O_{2+} + O_3$ $k_{148} = 3E - 16 \text{ m}^3 \text{ s}^{-1}$

149 $O_2 + O_{4+} \longrightarrow 2O_2 + O_{2+}$ $k_{149} = 0.02673 \exp(-5030/T_g) \text{ m}^3 \text{ s}^{-1}$

150 $O_- + O_2(a^1\Delta) \longrightarrow O + O_2$ $k_{150} = 3.3E - 17 \text{ m}^3 \text{ s}^{-1}$

151 $O_- + O_2 \longrightarrow O + O_{2-}$ $k_{151} = 1E - 20 \text{ m}^3 \text{ s}^{-1}$

152 $O_- + O_2(a^1\Delta) \longrightarrow O + O_2 + e$ $k_{152} = 2E - 16\exp(-15\,000/T_g) \text{ m}^3 \text{ s}^{-1}$

153 $O_- + 2O_2 \longrightarrow O_2 + O_{3-}$ $k_{153} = 3.3E - 40T_g^{-1}\text{m}^6\text{s}^{-1}$

154 $O_{2-} + O_2 \longrightarrow 2O_2 + e$ $k_{154} = 2E - 16\exp(-5338/T_g) \text{ m}^3 \text{ s}^{-1}$

155 $O_- + O_+ \longrightarrow O_2$ $k_{155} = 2.7E - 13 \text{ m}^3 \text{ s}^{-1}$

156 $O_- + O_{4+} \longrightarrow O_2 + O_3$ $k_{156} = 6.9E - 12T_g^{-0.5}\text{m}^6\text{s}^{-1}$

157 $O_- + O_2 + O_{2+} \longrightarrow O_2 + O_3$ $k_{157} = 2E - 37 \text{ m}^6 \text{ s}^{-1}$

(Continued)

Table 8.4. (*Continued*)

No. Reaction	$\Delta\varepsilon$, eV Rate constant
158 $O_{2-} + O_{2+} + O_2 \longrightarrow 3O_2$	$k_{158} = 2E - 37 \text{ m}^6 \text{ s}^{-1}$
159 $O_{2-} + O_{4+} \longrightarrow 3O_2$	$k_{159} = 1E - 13 \text{ m}^3 \text{ s}^{-1}$
160 $O + O_{3-} \longrightarrow 2O_2 + e$	$k_{160} = 3E - 16 \text{ m}^3 \text{ s}^{-1}$
161 $O_2 + O_{3-} \longrightarrow 2O_2 + O_-$	$k_{161} = 1.62E - 6T_g^{-2} \exp(-18\,260/T_g) \text{ m}^6 \text{ s}^{-1}$
162 $O_{3-} + O_{4+} \longrightarrow 3O_2 + O$	$k_{162} = 1E - 13 \text{ m}^3 \text{ s}^{-1}$
163 $O_2 + O_{2-} + O_{4+} \longrightarrow 2O_2$	$k_{163} = 4E - 38 \text{ m}^6 \text{ s}^{-1}$
164 $O_2 + O_{3-} + O_{4+} \longrightarrow 2O_2 + O$	$k_{164} = 4E - 38 \text{ m}^6 \text{ s}^{-1}$

whereas in the kinetic model, they were obtained by convoluting these cross-sections with a calculated EDF.

It can be seen from figures 8.17, 8.18 and 8.19 that the spatial distribution of the charged particle densities is highly non-uniform over the discharge cross-section. Almost all of the negative ions reside in the inner ion–ion plasma region. The radius of this region is $r = r_0$. The external electron–ion plasma region ($r_0 < r < R_d$) consists of electrons and positive ions, whereas the negative ions are practically absent there.

Since the flux of negative ions to the wall is zero, we find from equation (8.27) that the densities averaged over the cross-section satisfy the relationship [14, 37]

$$\nu_a \bar{n}_e = \nu_d \bar{n}_n + K_r \bar{n}_n \bar{n}_p. \tag{8.30}$$

Dividing equations (8.26) and (8.27) by the corresponding diffusion coefficients and summing them we arrive at the equation [14, 36, 38]

$$-2\Delta n_n/k - \Delta n_e = n_e/l_e^2 - 2n_n/kl_n^2, \tag{8.31}$$

which is of fundamental importance for analysing the set of equations (8.26) and (8.27).

Equation (8.31) contains two characteristic space scales, l_e and l_n, which are defined by

$$1/l_e^2 = 1/l_{ion}^2 + 1/l_a^2 = \nu_i/D_{ap} + \nu_a/D_{an} = \tau_{ap}\nu_i/\Lambda^2 + \tau_{an}\nu_a/\Lambda^2, \tag{8.32}$$

$$1/l_n^2 = 1/l_{nd}^2 + 1/l_{nr}^2 = \nu_a/2D_n + n_p K_r/D_{np} = \tau_n \nu_d/\Lambda^2 + \tau_{np} n_p K_r/\Lambda^2 \tag{8.33}$$

where, $D_{an,ap} = D_{n,\,p}(k + 1)$ and $D_{np} = 2D_n D_p/(D_n + D_p)$ are the coefficients of electron–ion and ion–ion ambipolar diffusion, respectively; $\tau_j = \Lambda^2/D_j$ are the corresponding characteristic times; and Λ is the diffusion length, which, in the case of cylindrical geometry, is equal to $\Lambda = R_d/2.4$.

Since the ambipolar electric field draws negative ions into the plasma, their density in the external region ($r_0 \leqslant r \leqslant R_d$) is low, $n_n(r) \approx 0$; hence, we have $n_e(r) \approx n_p(r)$ in this region. At $k \gg 1$, we can write equation (8.31) in the form

$$\Delta n_e = n_e / l_e^2. \tag{8.34}$$

Taking into account the spread caused by ion diffusion, we find that the thickness of the external region satisfies the condition $R_d - r_0 \leqslant l_e$; i.e., l_e determines the maximum thickness of the shell. We would start with the case when this thickness is small compared with the tube radius R_d (and, hence, with the characteristic diffusion length $\Lambda = R_d / 2.4$). Therefore, the external region can be treated in plane geometry. Then, for the plasma density profile in the region $r_0 \leqslant r \leqslant R_d$, we can use the solution

$$n_e(r) = n_e(r_0) \sin(\pi (R_d - r)/2l_e) / sin(\pi (R_d - r_0)/2l_e). \tag{8.35}$$

The density profiles in the inner region depend substantially on the ratio between R_d and l_n (see equation (8.33)), i.e., between the radius and the distance a negative ion covers due to diffusion during its lifetime with respect to volume processes [14, 38]. At $\tau_{an} \nu_a > 1$, length l_e (8.32) is small ($l_e < \Lambda$), and, under typical discharge conditions ($\bar{n}_n / \bar{n}_e < k \approx 100$), the length l_n turns out to be even smaller ($l_n < l_e$); hence, ion diffusion can be ignored [34–36]. When the opposite inequality is satisfied ($\tau_{an} \nu_a < 1$), the electron–ion plasma occupies almost the entire cross-section of the tube, whereas the length l_n can be either longer or shorter than the radius of the inner ion–ion region. Hence, to obtain functional dependences in the inner region, it is reasonable to consider two limiting regimes with large and small values of the parameter $\tau_{an} \nu_a$ which is quadratic in pressure.

For oxygen, the boundary value $\tau_{an} \nu_a = 1$ corresponds to $p\Lambda \approx 0.07$ cm Torr, so that $\tau_{an} \nu_a > 1$ at $p\Lambda > 0.07$ cm Torr and vice versa. Consequently, length l_e (equation (8.32)) has two asymptotes: $l_e \approx \Lambda$ at low pressures, $p\Lambda < 0.07$ cm Torr, and $l_e \approx l_n$ in the opposite case.

At high attachment frequencies ($\tau_{an} \nu_a > 1$), characteristic lengths, equations (8.32) and (8.33), as was mentioned above, are both small ($l_n < l_e < \Lambda$). Since $l_n < l_e$, we can neglect ion diffusion in equations (8.26) and (8.27) (as was done in [34–36]) and assume that the shell thickness is $R_d - r_0 \approx l_e$ (i.e., the denominator in equation (8.34) is equal to unity). In the case in hand, in balance equation (8.26) for negative ions, their transport is insignificant as compared to volume processes (see figure 8.20), so that the negative ion flux is almost completely determined by the drift component. Hence, at $n_p \approx n_n > n_e$, the fluxes of positive and negative ions in the inner region are almost the same in magnitude, but opposite in sign; i.e., we have [14, 36]

$$\Gamma_n / b_n \approx k n_p \nabla n_e / n_e \approx k n_n \nabla n_e / n_e \approx -\Gamma_p / b_p. \tag{8.36}$$

For this reason, in equation (8.31), in which these fluxes are summed, they almost completely cancel each other in the inner ion–ion region. In other words, at $r < r_0$, the terms on the left-hand side of equation (8.31) (which are responsible for spatial transport) are small compared to the terms on the right-hand side (which are

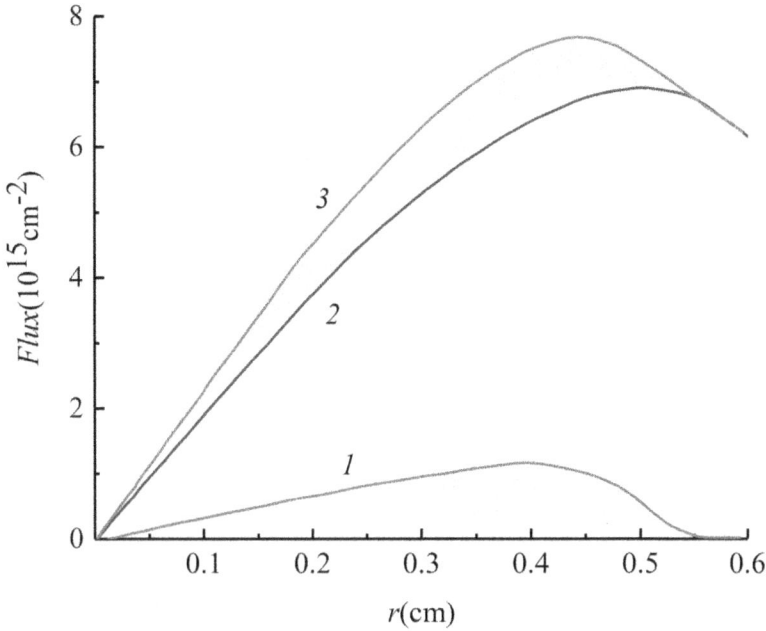

Figure 8.20. Contributions of spatial transport and volume processes to the negative ion balance for $p = 1$ Torr and $I = 50$ mA. Curve 1 shows the flux of negative ions (with a minus sign), curve 2 shows the total production of ions, and curve 3 shows the ion loss. From [10].

responsible for volume processes). Hence, the local balance of the volume plasma-chemical processes resulting in the production and loss of ions, $n_e/l_e^2 = 2n_n/kl_n^2$, holds with a high accuracy. At $\tau_{an}\nu_a > 1$, the important relation can be deduced from this equality [10, 14, 36]

$$(\nu_i/D_p + \nu_a/D_n)n_e = \nu_d n_n/D_n + K_r n_n(n_n + n_e)/(1/D_p + 1/D_n), \tag{8.37}$$

which allows one to obtain the relationships between the plasma parameters in the central region $r < r_0$.

The relationships between the densities of charged particles depend on the mechanism responsible for the loss of negative ions, i.e., on the relationship between the terms on the right-hand side of equation (8.37). At $\tau_{an}\nu_a > 1$, the loss of negative ions in an oxygen plasma is governed by detachment processes (the detachment regime with $\nu_d > n_p K_r$). Then, it follows from equation (8.37) that the profiles of the electron and negative ion densities are similar

$$\nabla n_e/n_e = \nabla n_n/n_n, \quad n_e(x)/n_n(x) = \text{const.} \tag{8.38}$$

This condition was first proposed in [39] and then was justified in [34–36], assuming that ion diffusion can be neglected as compared to ion drift. It follows from the above analysis that equation (8.37) is valid only at $\tau_{an}\nu_a > 1$; it is impossible to extrapolate it to the low pressure.

The validity of equation (8.38) for oxygen is illustrated in figure 8.21, which shows the density profiles from figure 8.18 ($p = 1$ Torr) normalized to the central electron density. Substituting equation (8.38) into equation (8.26) or (8.27), we find that, at a significant degree of electronegativity ($n_n > n_e$), the densities in the inner region are

$$n_p(r) \sim n_n(r) \sim n_e(r) \sim J_0(r/l_0). \tag{8.39}$$

For plane geometry, the Bessel function should be replaced with $\cos(x/l_0)$. In equation (8.39), the characteristic length [14, 38]

$$l_0^2 = 2D_{an}/\nu_d + \nu_a D_{ap}/(\nu_i \nu_d) \approx \Lambda^2 n_n/(\nu_i \tau_{ap} n_e) > \Lambda^2 \tag{8.40}$$

also determines the ambipolar electric field ($E(r) = -T_e \nabla n_e/n_e$) in the central region ($r < r_0$):

$$E_c(r) \approx -T_e J_1(r/l_0)/l_0 \sim -T_e r/l_0^2. \tag{8.41}$$

At $l_0 > \Lambda > l_e$ the density profiles (8.39) in the inner region are flatter than in the external region and, being extended up to the wall, they do not turn to zero (see equation (8.29)). Consequently, the field (8.41) is weaker than the electric field in the shell ($r_0 < r < R_d$), for which we have from equation (8.35) the following estimate

$$E_s(r) \approx -(\pi T_e 2 l_e)\cot(\pi(R_d - r)/2l_e) \sim -(\pi^2 T_e/2l_e^2)(r - r_0). \tag{8.42}$$

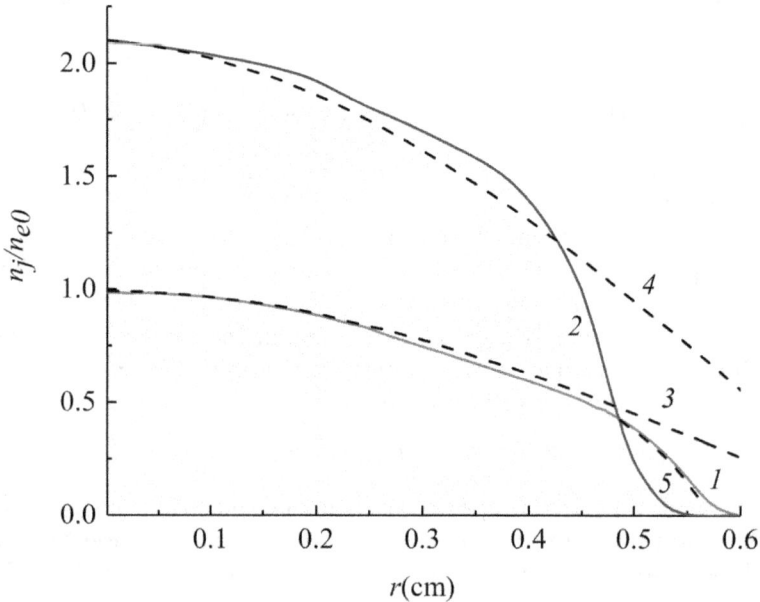

Figure 8.21. Normalized density profiles for $p = 1$ Torr and $I = 50$ mA: (1) $n_e(r)/n_{e0}$, (2) $n_n(r)/n_{e0}$. Curves 3 and 4 show the results calculated by equation (7.20), and curve 5 shows profile (7.17) in the outer region. From [10].

To illustrate the limiting cases, we use equations (8.26) and (8.27) to rewrite relationship (8.36) in the form [14, 36]

$$\Gamma_n = \nu_a \int_{r_0}^{R_d} n_e(r)r\,dr = -\Gamma_p D = D\nu_i \int_0^{r_0} n_e(r)r\,dr \qquad (8.43)$$

where $D = (D_n/D_p) \sim 1$. This relationship means that the attachment rate in the external region is equal to the number of ions produced in the central region due to ionization. In the thin shell, a comparatively small flux of negative ions Γ_n is produced due to attachment; hence, a fairly weak electric field (8.41) is sufficient to transport these ions into the inner region, in which they disappear due to detachment. Since, at $\tau_{an}\nu_a > 1$, the electrons in the inner region disappear mainly due to attachment, it is necessary to enable just a minor flux of positive ions toward the external region. In other words, relationship (8.43) means that, if the local plasma-chemical balance of ions dominates over their spatial transport, the latter should only compensate for a relatively small difference between the attachment and detachment of negative ions.

Using expression (8.35) for $n_e(r)$ and equations (8.41) and (8.42), we can obtain from equation (8.43) the ionization frequency Z_i, which represents the eigenvalue of the boundary value problem described by equations (8.26) and (8.27) [14]. The simple estimate $\Gamma_n \approx \nu_a n_e l_e \approx \Gamma_p \approx Z_i n_e \Lambda$ gives $Z_i \approx \nu_a l_e/\Lambda \approx \sqrt{\nu_a/\tau_{an}}$ [14, 36].

In the case at hand, we have $\nu_i \tau_{an} \approx \sqrt{\nu_a \tau_{an}} > 1$; hence, we obtain $\tau_{an} Z_i > 1$. This means that the ionization frequency exceeds the value given by the Schottky formula for a simple plasma ($\tau_{an} Z_i = 1$) [14, 36].

The small density of the negative ions that are produced in the shell due to attachment can be deduced from their flux Γ_n (8.43)

$$n_n \approx \Gamma_n/b_n E_s \approx 8l_e^2 n_e(r_0)/(\pi^2 D_{an})\sin^2(\pi(R_d - r)/2l_e)\tan(\pi(R_d - r)/2l_e) \qquad (8.44)$$

$$\approx \pi n_e(r_0)\nu_a(R_d - r)^3/(4D_{an}l_e) \quad (r_0 < r \leqslant R_d). \qquad (8.45)$$

At the point $r = r_0 \approx R_d - l_e$, the field E_s is close to zero, whereas the flux Γ_n (8.42) caused by attachment in the external region, is finite. Therefore, when approaching the point $r = r_0$, negative ion density (equation (8.45)) sharply increases

$$n_n \approx 8l_e^3 2n_e(r_0)\nu_a/(\pi^3 D_{an}(r - r_0)) \quad (r \geqslant r_0 = R_d - l_e) \qquad (8.46)$$

to its value in the inner region, which is determined by equation (8.37). The transition zone separating regions with different ion composition is narrow ($\sim l_n < l_e$). For this reason, it was treated in [14, 36], as a diffusive jump in which ion densities undergo a jump, whereas the ion fluxes and the electron density are continuous. The validity of relationship (8.37) in the region $r < r_0$ in an oxygen discharge is illustrated in figure 8.21, which shows the normalized density profiles from figure 8.17 at a pressure of $p = 1$ Torr. The dashed curves in figure 8.21 show the profiles calculated by formula (8.39) for the inner region and by formula (8.35) for the shell with the thickness $R_d - r_0 \approx l_e$. When deducing formula (8.35) for the external region, the shell thickness

δ_{sh} was taken into account; i.e., it was assumed that the electron density vanished at $r = R_d - \delta_{\text{sh}}$, rather than at the tube wall. It can be seen that the results of calculations using these formulas agree well with the results of full-scale simulations.

At lower pressures, the role of spatial transport increases and, thus, the characteristic lengths l_e (8.32) and l_n (8.33) also increase. The increase in length l_e (8.32) leads to the flattening of density profiles (8.39) in the inner region. Because of the increase in length l_n, the region with a sharp change of the ion density spreads out due to ion diffusion; hence, the transient region can no longer be treated as a jump. As a result, the ion density profiles become bell-shaped.

At $l_n \geqslant \Lambda$, the negative ions are able to pass throughout the entire discharge volume due to their diffusion. However, they remain trapped in the inner region by the electric field; as a result, a Boltzmann distribution (similar to that for electrons) is established

$$-T_e \nabla n_e / n_e = -T \nabla n_n / n_n = E. \tag{8.47}$$

It follows from equations (8.32) and (8.33) that, generally, the own diffusion of the negative ions prevails ($l_n > \Lambda$) only when attachment is insignificant as compared to the ambipolar diffusion of negative ions (ion diffusion with the electron temperature), i.e., when $\tau_{an}\nu_a \ll 1$ (see [14, 10] for details).

Condition (8.47) leads to the relationship

$$n_e(r)/n_e(0) = [n_n(r)/n_n(0)]^{1/k}, \tag{8.48}$$

which strongly depends on the temperature ratio $k = T_e/T_i$ and coincides with distribution (8.38) only in the particular case $T_e = T_i$. The establishment of a Boltzmann distribution for electrons and negative ions at low pressures is illustrated in figure 8.22, in which the simulation results shown in figure 8.19 for a pressure of $p = 0.15$ Torr are replotted in accordance with equation (8.48)[1].

Since $k \gg 1$ in discharges, it follows from equation (8.48) that the electron density profile is nearly flat, $n_e(r) \sim n_{e0} \approx$ const, which is indeed observed at reduced pressures (see figures 8.18 and 8.19). Here, transport processes play a major role for the negative ion balance (see figure 8.23), in contrast to the above case with $\tau_{an}\nu_a > 1$ (cf figure 8.20). The field-induced and diffusive fluxes of negative ions are almost the same in magnitude, but opposite in sign; hence, a small difference between them is sufficient to balance the production and loss of ions at any point (figure 8.23). The plasma-chemical processes govern only the global balance of ions in the central region. In equation (8.26) for the positive ion density $n_p(x)$, the terms on the left-hand side are also approximately equal to each other. However, they are summed and, thus, at a significant degree of electronegativity ($n_n(0) > n_e(0)$), balance equation (8.26) for positive ions can be written in the form $-2D_p\Delta\, n_n = Z_i n_{e0}$. This

---

[1] Note that, for the recombination regime ($\nu_d < n_p K_r$), it follows from equations (8.36) and (8.37) that $\nabla n_e/n_e = \nabla n_n/n_n + \nabla n_p/n_p \approx 2\nabla n_n/n_n$, which results, in contrast to equation (8.38), in an ion distribution that is more flat than the electron distribution (see [10, 14] for details). In such a situation (which occurs, e.g. for halogens), the attachment and ionization frequencies are approximately the same, $Z_i \approx \nu_a$, as was noted in [35].

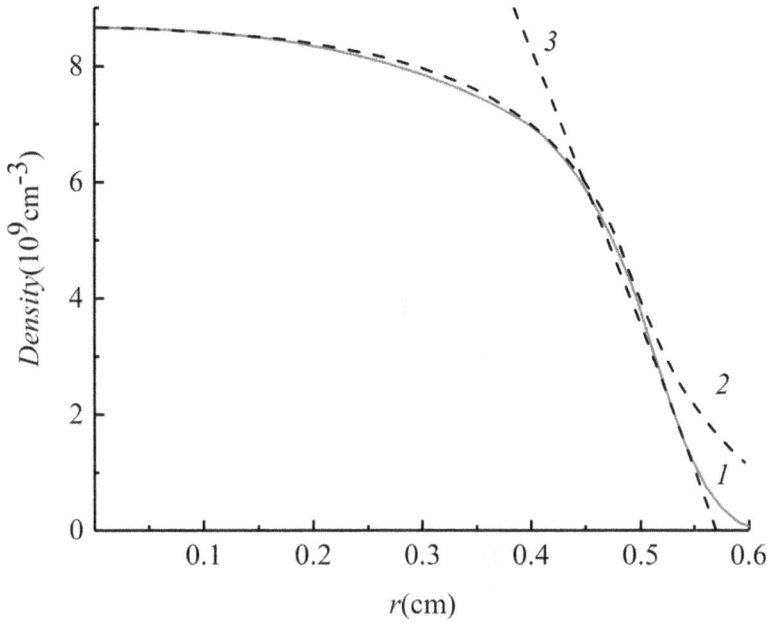

Figure 8.22. Boltzmann distributions of electrons and negative ions for $p = 0.15$ Torr and $I = 50$ mA: (1) $n_e(r)$, (2) $n_e(0)/[n_n(r)/n_n(0)]^{1/k}$ (see equation (8.48)), and (3) electron density profile (8.35) in the outer region. From [10].

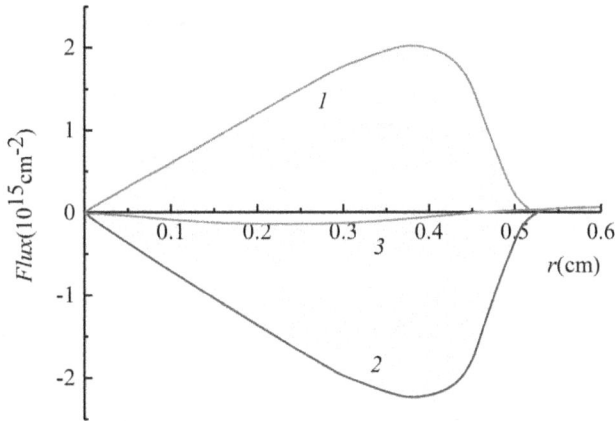

Figure 8.23. Contributions of spatial transport and volume processes to the negative ion balance for $p = 0.15$ torr and $I = 50$ mA: (1) the diffusion component of the negative ion flux, (2) its drift component, and (3) the resulting flux balancing the production and loss of negative ions in volume plasma-chemical processes. From [10].

gives a parabolic distribution of the ion densities and a flat profile of the electron density $n_e(r)$ at $r < r_0$ [33]:

$$n_n(r) = n_{n0}[1 - r^2/r_0^2], \quad n_{n0}/n_{e0} = \nu_i r_0^2/4D_p, \quad n_e(r) \approx n_{e0} \approx \text{const.} \quad (8.49)$$

We note that the ion diffusion in the inner region proceeds with the coefficient $2D_p$ of the own ion–ion ambipolar diffusion, rather than with the usual coefficient of ambipolar diffusion $D_p(1 + k)$. It can be seen from figures 8.18 and 8.19 that, at low pressures, simple parabolic law (8.49) for the ion density profiles agrees well with the results of full-scale simulations[2].

In the external region (shell), in which the negative ions are almost absent, the plasma density profile varies in accordance with equation (8.35). In [10, 33], the position of the boundary point $r = r_0$ was found from the negative ion balance using model profiles (8.49). Unfortunately, this procedure is rather laborious and provides a low accuracy.

It seems that the position of the boundary can be found in a more simple and reliable way from the continuity of the positive ion flux at $r = r_0$

$$2D_p n_{n0}/r_0 = D_p(1 + k)n_{e0}/(l_e \tan(R_d - r_0)/l_e)) \approx D_p(1 + k)n_{e0}/(R_d - r_0). \quad (8.50)$$

Model electron density profiles (8.35) with r_0 defined by equation (8.50) (see figure 8.22) agree well with the results of full-scale simulations shown in figure 8.19.

Based on the analysis performed, we recommend the following procedure to obtain approximate density profiles in the plasma of electronegative gases in the detachment regime ($\nu_d < K_r n_p$):

(i) First, the parameter $\tau_{an}\nu_a$ is estimated.

(ii) Then it is necessary to indent from the wall by the thickness δ_{sh} of the space-charge sheath, which can be estimated, e.g. according to [14].

(iii) In the external electron–ion plasma region ($r_0 < r < R_d$), where $n_p \approx n_e \gg n_n \approx 0$, the electron density varies according to equation (8.35) and the negative ion density varies according to equation (8.46). If $\tau_{an}\nu_a > 1$, then the thickness of this region is equal to l_e (see equation (8.32)) and the denominator in equation (8.35) is equal to unity ($r_0 = R_d - l_e$). In the opposite case ($\tau_{an}\nu_a < 1$), we have $l_e \approx \Lambda$ and the thickness of this region is estimated by formula (8.50).

(iv) Finally, the density profiles in the central region ($r < r_0$) are determined.

At $\tau_{an}\nu_a > 1$, the density profiles are similar and are described by equation (8.39), whereas the density values are related by expression (8.37). Electron density profile (8.39) is matched to expression (8.35) at $r = r_0 = R_d - l_e$. The ion densities undergo a jump at this point: the negative ion density drops to nearly zero (see equation (8.45)), whereas the positive ion density decreases to the value equal to the electron density given by equation (8.35). At $\tau_{an}\nu_a > 1$, the thickness of the transition zone ($\sim l_n < l_e$) is small and it can be regarded as a jump in the ion density.

At $\tau_{an}\nu_a < 1$, the electron density profile is flat ($n_e(x) \approx n_{e0}$) and the ion density profile is parabolic. These densities are related by formulas (8.49). The electron

---

[2] We note also that, in order for profiles (8.49) to be established, it is enough to satisfy the condition $\tau_{an}\nu_a < 1$. The mechanism for the volume loss of negative ions, which is determined by the right-hand side of equation (8.27), can be either recombination (at $\nu_d < K_r n_p$) or detachment (at $\nu_d > K_r n_p$).

density profile is matched at the point $r = r_0$, whose position can be estimated from equation (8.50).

If the ion diffusion can be negligible in this case, the peripheral region of the pure (electron–ion) plasma occupies practically the whole column cross-section, i.e., the central region is small, $r_0 \ll R_d$ [34–36]. In this situation the condition of discharge maintenance coincides with the condition for pure plasma (8.2) both in the cases of the detachment-dominated and of the recombination-dominated plasma core cases. It follows from the fact that the main part of the ionization, as well, as of the diffusion, takes place in the pure electron–ion plasma, and the processes in the relatively small ion–ion central core give a negligible contribution in the overall balance of the positive ions.

Substituting the calculated expressions for ν_{ion}, and using the shock condition (8.36) together with the overall balance of negative particles, (8.30), and with the partial density profiles (8.35) and (8.39), it is easy to find the coefficients in equations (8.35), (8.39) and the ion density jump at the shock.

The width l_1 of the central ion–ion plasma core equals

$$l_1 = 2L\nu_a\tau_{an}D/(\pi D_n). \tag{8.51}$$

The central density of the negative ions is

$$n_0 = n_{e0}D_n/(\nu_d\tau_{an}D_p) \tag{8.52}$$

It should be noted that the spatially averaged electronegativity $\bar{n}_n/\bar{n}_e$ can be rather large in this case.

Since the electron attachment results in considerable reduction of the ionization rate (n_eZ_i), for plasma maintainence of the strongly electronegative gases rather than high reduced field (E/p) are necessary. In the fluid ion description used it means that the ion temperature can increase significantly, particularly, at low pressures [40]. As the pressure decreases, the directed velocity acquired by the ions in the field can become higher than the random (thermal) velocity. The coefficient of ion diffusion also increases. This can dramatically change the ion density profiles [40], presented in figures 8.18 and 8.19, and shows that taking into account ion heating (which increases the ion diffusion coefficient) dramatically changes the shell thickness. For this reason, when analysing the spatial profiles of the charged particle densities in electronegative gases, one of the central problems is the ion temperature [40].

The ion density profiles computed for $p = 0.15$ Torr without and with allowance for ion heating in the longitudinal electric field are shown in figures 8.18 and 8.19, respectively.

Ion heating was calculated by the formulas for the effective transverse ion temperature [40]

$$T_i = T + \frac{(M_i + M)Mw^2}{3(2M + M_i)}, \tag{8.53}$$

where M and M_i are the masses of a molecule and an ion, respectively, and w is the ion drift velocity in the longitudinal electric field E_z. For example, at $p = 1$ Torr, the transverse ion temperature is ≈ 760 K, whereas at $p = 0.15$ Torr, it is ≈ 5200 K. For oxygen, an order of magnitude of the ion temperature as a function of the parameter $p\Lambda$ is presented, e.g. in [40].

In [38] the kinetic and fluid approaches were compared to the modeling of the molecular plasma of the positive column of a dc oxygen discharge in a 12 mm diameter glass tube at gas pressures of 0.5–3 Torr and discharge currents of 5–200 mA.

As oxygen is a molecular gas, for the correct description of the glow discharge it is necessary to consider big enough nomenclature plasma-chemical processes with the participation of various atomic and molecular oxygen states. The accounted set are listed in table 8.4.

In the fluid model, the rate constants of the processes with the participation of electrons were obtained by convoluting the corresponding cross-sections with a Maxwellian EDF, whereas in the kinetic model, they were obtained by convoluting these cross-sections with an EDF calculated with the help of the Comsol Multiphysics [13].

Typical EDFs obtained by self-consistently simulating the dc discharge plasma at gas pressures of $p = 1$ and 0.15 Torr are shown in figures 8.24 and 8.25 It can be seen that all the EDFs are strongly non-equilibrium. For this reason, the rate constants of many plasma-chemical processes differ significantly from those obtained in the fluid

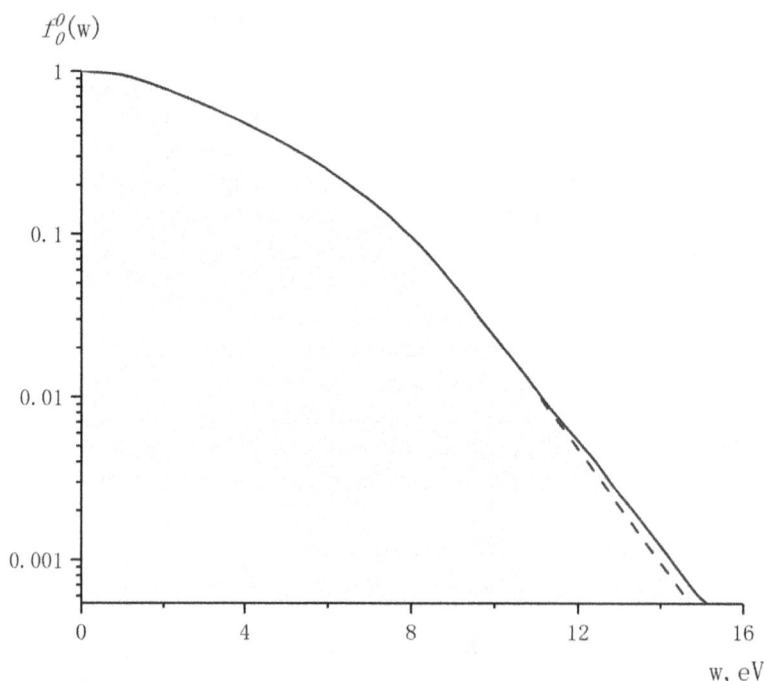

Figure 8.24. Local EDFs for $p = 1$ Torr and $i = 50$ mA. The solid curves shows the results of self-consistent calculations for the radius r varying from 0 to R_d with a step of $R_d/5$. The dashed curve shows the local EDF $f_0^0(w)$. From [38].

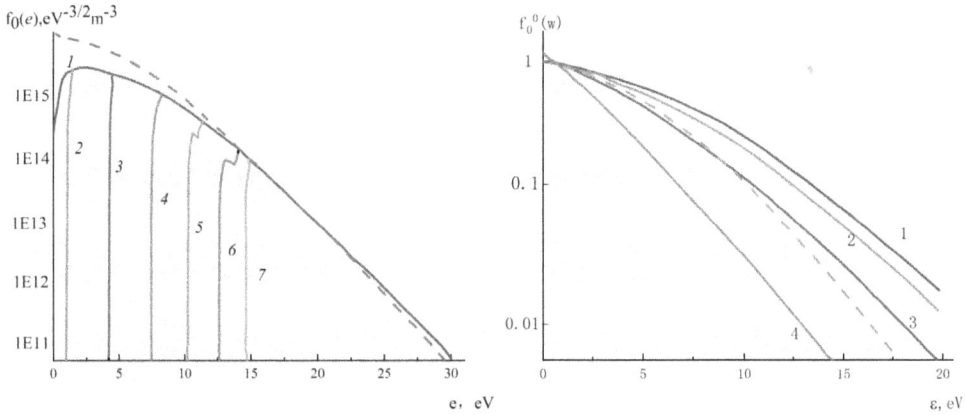

Figure 8.25. Non-local EDFs for $p = 0.15$ Torr and $i = 50$ mA. (a) The solid curves show the results of self-consistent calculations for the radii $r = $ (1) 0, (2) $0.6R_d$, (3) $0.8R_d$, (4) $0.69R_d$, (5) $0.95R_d$, (6) $0.98R_d$, and (7) R_d. The dashed curve shows the local EDF $f_0^0(w)$. (b) The solid curves show the results of self-consistent calculations for the radii $r = $ (1) 0, (2) $0.6R_d$, (3) $0.8R_d$, and (4) R_d. The dashed curve shows the local EDF $f_0^0(w)$. From [38].

model, which deals with the same temperature all the electrons. In turn, this leads to the difference in the plasma parameters that significantly depend on these rate constants. At high pressures, the EDF is close to local; i.e. the function $f_0^0(w)$ is independent of r (figure 8.24). At low pressures, the EDF is not only non-equilibrium but also non-local. In this case, the values of $f_0(\epsilon)$ represented as a function of the total energy $\epsilon = w + e\phi$ (r) (i.e., without normalizing and shifting by the space potential) coincide at different radii (figure 8.24a). The same EDFs $f_0(\epsilon)$ plotted using the conventional local representation as functions of the kinetic energy w (similar to figure 8.24) differ for different radii r (figure 8.24b).

In oxygen, the energy loss due to inelastic collisions is dominant over almost the entire energy range. Hence, using the equality $\lambda_\epsilon = \sqrt{\lambda}\lambda^*$ and the total cross-sections for elastic and inelastic collisions ($\sigma = 5 \times 10^{-16}$ cm^2 and $\sigma^* = 5 \times 10^{-17}$ cm^2, respectively), we obtain the estimate $\lambda_\epsilon \simeq 0.2/p$ (in cm), where p is in Torr. It follows from the above estimates that, for oxygen, the criterion for the EDF to be non-local ($\lambda_\epsilon > \Lambda$) is $p\Lambda < 0.2$ cm Torr, which agrees with the results of simulations [38].

References

[1] Raizer Y P 1991 *Gas Discharge Physics* (Berlin: Springer)
[2] Golubovskii Y B, Kudryavtsev A A, Nekuchaev V O, Porohova I A and Tsendin L D 2004 *Electron Kinetics in Non-equilibrium Gas-discharge Plasma* (St Petersburg: SPbSU) [in Russian]
[3] Langmuir I 1925 Scattering of electrons in ionized gases *Phys. Rev.* **26** 585
Langmuir I 1929 The interaction of electron and positive ion space charges in cathode sheaths *Phys. Rev.* **23** 954
Tonks L and Langmuir I 1929 The positive column of gas discharge *Phys. Rev.* **34** 876
[4] Schottky W 1924 Diffusionstheorie der positiven Saule *Z. Phys.* **25** 635

[5] Caruso A and Cavalieri A 1962 The structure of the collision-less plasma-sheath transition *Nuovo Cimento* **26** 1389

[6] Ingold J H 1978 *Gaseous Electronics* ed M N Hirsh and H J Oskamvol I (New York: Academic), p 36

[7] Kudryavtsev A A and Tsendin L D 1999 Mechanisms for formation of the electron distribution function in the positive column of discharges under Langmuir-paradox conditions *Tech. Phys.* **44** 1290–7

[8] Behnke J, Golubovskii Y B, Nisimov S U and Porokhova I A 1996 Self consistent model of a positive column in an inert gas discharge at low pressures and small currents *Contrib. Plasma Phys.* **36** 75

[9] Golubovsky Y B, Porokhova I A and Behnke J 1999 A comparison of kinetic and fluid models of the positive column of discharges in inert gases *Phys. D: Appl. Phys.* **32** 456–70

[10] Bogdanov E A, Kudryavtsev A A, Tsendin L D, Arslanbekov R R, Kolobov V I and Kudryavtsev V V 2003 Scaling laws for the spatial distributions of the plasma parameters in the positive column of a dc oxygen discharge *Tech. Phys.* **48** 1151–8

[11] Bogdanov E A, Kudryavtsev A A, Tsendin L D, Arslanbekov R R, Kolobov V I and Kudryavtsev V V 2004 The influence of metastable atoms and the effect of the nonlocal character of the electron distribution on the characteristics of the positive column in an argon discharge *Tech. Phys.* **49** 698–706

[12] Bogdanov E A, Kudryavtsev A A, Tsendin L D, Arslanbekov R R and Kolobov V I 2004 Nonlocal phenomena in the positive column of a medium-pressure glow discharge *Tech. Phys.* **49** 849–57

[13] http://www.comsol.com

[14] Rozhansky A V and Tsendin L D 2001 *Transport Phenomena in Partially Ionized Plasma* (London: Taylor & Francis)

[15] Tsendin L D 1995 Electron kinetics in non-uniform glow discharge plasmas *Plasma Sources Sci. Technol.* **4** 200

[16] Compton T and Langmuir I 1930 Electrical discharges in gases. Part I. Survey of fundamental processes *Rev. Mod. Phys.* **2** 123
Compton T and Langmuir I Electrical discharges in gases. Part II. Fundamental phenomena in electrical discharges *Rev. Mod. Phys.* **3** 191

[17] Gabor D, Ash E A and Dracott D 1955 Langmuir's paradox *Nature* **176** 916–9

[18] Chen F 2016 *Introduction to Plasma Physics and Controlled Fusion* (New York: Springer)

[19] Benke Yu, Kagan Yu M and Milenin V M 1969 On Langmuir paradox *Sov. Phys. Tech. Phys.* **13** 989
Kagan Yu M 1965 On electron velocity distribution function in a discharge positive column *Beitr. Plasma Phys.* **5** 479 [in Russian]

[20] Kagan Y M 1970 Electron velocity distribution function in a discharge positive column *Spectroscopy of Gas Discharge Plasmas* (Leningrad: Nauka), 201–23 [in Russian]

[21] Milenin V M and Timofeev N A 1978 Electron relaxation to a Maxwellian energy distribution in the positive column of a low-pressure mercury discharge *Sov. Phys. Tech. Phys.* **23** 1048

[22] Rompe R, Ullrich S and Wolf H 1960 Zum Verhalten von Elektronen in Niederdruckplasmen *Beitr. Plasmaphys.* **1** 245–9

[23] Hibsch E H 1966 Plasma probes and the Langmuir paradox *Int. J. Electron.* **19** 537–41

[24] Von Gierke G, Ott W and Schwirzke F 1961 *Proc. 6th ICPIG, Munchen* vol 11, p 1412

[25] Ott W 1963 Ein Versuch zur Klarung des Langmuir-Paradoxus Preprint Inst. Plasmaphys., Munchen, IPP 2/19 27 S

[26] Harp R and Kino G S 1963 *Proc. 8th IGPIG Paris* vol 3, p 45

[27] Crawford F W and Self S A 1965 On the low-pressure mercury-vapor discharge mechanism and the origins of Langmuir's paradox *Int. J. Electron.* **18** 569–77

[28] Rayment S W and Twiddy N D 1968 Electron energy distributions in the low-pressure mercury-vapour discharge: the Langmuir paradox *Proc. Roy. Soc.* A **340** 87–98

[29] Baalrud S D, Callen J D and Hegna C C 2009 Instability-enhanced collisional effects and Langmuir's paradox *Phys. Rev. Lett.* **102** 245005

[30] Tsendin L D and Golubovsky Y B 1977 Positive column of a low-density, low-pressure discharge. I-Electron energy distribution *Sov. Phys. - Tech. Phys.* **22** 1066

[31] Kortshagen U, Parker G J and Lawler J E 1996 Comparison of Monte Carlo simulations and nonlocal calculations of the electron distribution function in a positive column plasma *Phys. Rev.* E **54** 6746

[32] Godyak V A and Alexandrovich B 2015 Langmuir paradox revisited *Plasma Sources Sci. Technol.* **24** 052001

[33] Lichtenberg A J, Vahedi V and Lieberman M A 1994 Modeling electronegative plasma discharges *J. Appl. Phys.* **75** 2339

[34] Ferreira C M, Gousset G and Touzeau M 1988 Quasi-neutral theory of positive columns in electronegative gases *J. Phys.* D **21** 1403

[35] Daniels P R and Franklin R N 1989 The positive column in electronegative gases-a boundary layer approach *J. Phys.* D **22** 780
Daniels P R, Franklin R N and Snell J 1990 The contracted positive column in electro negative gases *J. Phys. D: Appl. Phys.* **23** 823
1993 Characteristics of electric discharges in the halogens: the recombination-dominated positive I column *J. Phys. D: Appl. Phys.* **26** 1638–49

[36] Tsendin L D 1985 Diffusion of charged particles in the plasma of electronegative gases *Sov. Phys. Tech. Phys.* **30** 1377
Tsendin L D 1989 Plasma stratification in a discharge in an electronegative gas *Sov. Phys. Tech. Phys.* **34** 11

[37] Lieberman M and Lichtenberg A 2005 *Principles of Plasma Discharges and Materials Processing* (New York: Wiley)

[38] Bogdanov E A, Kudryavtsev A A, Tsendin L D, Arslanbekov R R, Kolobov V I and Kudryavtsev V V 2003 Substantiation of the two-temperature kinetic model by comparing calculations within the kinetic and fluid models of the positive column plasma of a dc oxygen discharge *Tech. Phys.* **48** 983–94

[39] Oskam H J 1958 Microwave investigation of disintegrating gaseous discharge plasmas *Philips Res. Rep.* **13** 335–57

[40] Ivanov V V, Klopovsky K S, Lopaev D V, Rakhimov V V and Rakhimova T V 1999 Experimental and theoretical investigation of oxygen glow discharge structure at low pressures *IEEE Trans. Plasma Sci.* **27** 1279